ETHICS FOR THE PROFESSIONAL SURVEYOR

A Collection of Thoughts

ETHICS FOR THE PROFESSIONAL SURVEYOR
A Collection of Thoughts

by

Dennis J. Mouland, PS

Landmark Enterprises
10324 Newton Way
Rancho Cordova, CA 95670

ISBN: 0-910845-57-3

Dedication

To my wife and children, who still wonder
why I get so excited about
old marks on trees and rocks.

TABLE OF CONTENTS

Table of contents continued

Appendices

INTRODUCTION

We land surveyors are a curious group.......

Any surveyor worth his or her salt has a good library in the office. It would typically be loaded with books on the technical side of surveying. There would also be (hopefully) a number of books on the subject of legal principles. Recent new books in the surveying area are usually rehashes of the same old principles. As Roy Minnick recently wrote while reviewing a new book, "everything that can be said about land descriptions has been said a dozen times at least". And I know that to be true. I wrote some on that subject for a "new" surveying book too!

Obviously as new technologies emerge, books can be written on a number of "new" areas, such as GPS, or GIS and how these relate to the professional surveying world. As surveying becomes more and more a highly specialized field, books are being written to help the non-surveyor in a real estate related field understand just what we do for a living. But the one subject that has never been written on in detail is that of ethics for the professional land surveyor.

Ethics? I've known many members of other professions who felt surveyors had no ethics! Some of our own profession feel ethics to be a burden to their employment potential! Ethics is not an easy subject to write about, or talk about, or change. The western culture resists any hint of somebody telling a "professional" what to do or how to do it. Human nature does not like rules. As they say, rules are made to be broken, and surveyors have taken opportunity to break every rule I know of. I've done a lot of thinking about the subject, I've interviewed hundreds of surveyors on the subject, and I've even lectured on the subject at a few conventions and meetings. So I'm not going to give you a set of rules. Rather, I see the need for surveyors to find common ground, to establish certain basic criteria on the principles that are unique to our profession. And the only way to "get away with" giving you some hidden rules, is to lay out some true anecdotes to help you think about these issues.

Are ethics important? Well ask Jim Baker, or Jimmy Swaggert, or Richard Nixon, or Gary Hart. Ask Mike Espy, who flew for free in an airplane owned by Tyson Chicken. You bet they are important, but often times people do not pay attention to them until they have gone past someone else's idea of "the edge".

Ethical principles are at the very heart of our profession. While doing a seminar in Arkansas, a young man made a comment to the entire group that really stayed with me. "Our business is ethics." Just as those listed above paid dearly for their lack of attention to ethical values, so will anyone in our profession. The penalties can come financially, personally, or relationaly. But one way or another, ethics must be an integral part of the surveyor's lifestyle.

I reviewed a number of books on ethics for other professions. One for engineers was written just like an engineer would write it! It was filled with ethical equations: If this happens, then do that. It sounded more like an algebraic expression than an ethical discussion. If x, then y. I could not produce such a book for my chosen profession. I wanted something thought-provoking, but not in the traditional textbook style. So for those of you looking for "cookbook" solutions to the hard ethical issues facing the modern surveyor, you may be disappointed. This book is more a friendly chat with my fellow professionals about an important but difficult subject.

There have been some excellent articles over the years in various surveying publications regarding the subject of ethics. In particular, Robert Foster, Peter Dale, and Jerry Broadus (through his P.O.B. articles) have provided the bulk of our ethics input the past few years. This book will in no way detract from these works. Rather, I hope it simply adds to the body of knowledge and thought-provocation.

So please take this book as some friendly thoughts and ideas to help you start a regular habit of analyzing your own ethics, and those of the other professionals around you. I promise not to preach too much (but I have done a little). I hope I can cause you to think deeply about this marvelous and exciting profession we have chosen. We have an inherent responsibility to the profession itself. Your ethics impact hundreds of other people. Let's live up to that responsibility.

Chapter 1

ETHICS

"Integrity is, fundamentally, the value we place on ourselves. It is our abillity to make and keep commitments to ourselves..
Stephen R. Covey, The Seven Habits of Highly Effective People

An ethical surveyor? Some would claim this to be an oxymoron. But I would strongly disagree! No matter where you are on the "ethical spectrum", you still have ethics. What are ethics anyway?

The study of the subject of ethics is called "deontology". This comes from a Greek word which literally means "the study of what is lacking or is needed." The term "ethics" itself is generally defined as "the study of human conduct with an emphasis on the determination of what is right or wrong". This definition would imply that someone must establish standards. Throughout recorded history, attempts have been made to establish standards. The most classic examples are from books such as the Bible, Koran, or the teachings of the Buddha. From these texts, we have some famous standards that most people in western civilization are aware of, primarily the Ten Commandments, and such.

It is important to understand the sources of our ethics. Each human being is operating under some sort of personal ethics "system". One could be easily tempted to try to separate his or her professional ethics from one's personal ethics. When you consider the realities of human nature you must also consider the fact that we are operating as separate and distinct entities. We must realize how much our professional ethics are influenced, steered, and guided by our personal ethics system. In fact, these two systems are essentially identical.

So it would be good before we go into anymore discussion to stop and think for a few moments, where do your personal ethics come from? Where do your personal values come from? Your value system? What principles do you operate under in your personal life and where did they come from? Let's examine some of the specific areas that most personal ethics come from.

Obviously your home would be a major source. What kind of home did you come from? I'm not here to pass judgment on your home life. Whether it was good, bad or indifferent, it has influenced who and what you are as a person and as a professional. Some psychologists say that 95% of all homes are dysfunctional. If that's true then I wonder how it is that small minority of 5% would be considered normal. But nevertheless, all of us have differing values that were in our homes. There are certain things that could have drastically affected how you look at the world, how you look at your life, how you look at other people.

What about things that occurred that you had no control over at that time? For instance, were you raised in a single parent family? Was it your mother or your father? What were the circumstances that caused them to be separated?

What part of the country were you raised in? Where did you grow up and what kind of a lifestyle was it? Think about it. Does the child growing up in the ghettos of New York City have the same value system as you, who may have grown up on the farm? While there may be some similarities in human nature, the types of problems or opportunities you had greatly influenced your value system. How did this background influence your overall value of life, other people, of input, of variety? All of these things are effected by the home environment.

Another factor in the home could be your economic status while growing up as a child. I'm not really certain I could say which would be better, to grow up rich or poor, or in between. The fact is that there are some of us in the profession that grew up in all three of those categories and we have differing viewpoints. Some people who have been raised in a very poor (or perhaps just frugal) environment are very, very slow to take on expenditures such as new technology, new software, or new equipment. Others in the same sort of a background but with different values would revolt against that line of reasoning. These would be extremely quick to spend money, almost to a fault. Either way, we have allowed our background in some way to effect our professional operation in this life.

Another area within the home that cross references into other areas is your religious background. I'm not going to say what religion would be best or whether a religious background would be good for you or not. There's certainly a lot of evidence on every side of that argument. Someone who was raised with the basic American Judeo-Christian value system, perhaps based on the Ten Commandments, is more likely to be aware of certain values and principles that apply when dealing with other people in perhaps honesty and being fair. Certainly the Judeo-Christian background does not have a corner on this market. But someone who had no background in that sort of value system, who grew up quite amorally may not see things the same way.

Whether you agree with them or not, those standards of conduct in life were built into the person. But how were they portrayed? Were they portrayed in a hypocritical way?

Many people that have been raised in very strict and devout religious homes have rebelled and rejected that way because of the intense hypocrisy they saw in those that claimed those standards.

But one could look at the other extreme, that of the former Soviet Union prior to their latest revolution. This was a nation that was officially atheistic and operated without any general principles or morals. Their government was extremely corrupt. Everyone was stealing from someone else. They never could openly confess and admit that they shot down a Korean jetliner! Even after three days, they blatantly denied any involvement. The whole world knew it was the truth before their own people did.

So there's much to be said on both sides of the argument. But the point of this discussion is where did *you* come from? What was your background and how was it presented to you? How did you react to it as you grew up, as you became a teenager and then on into early adult life? What did you do with those values? What became of them? Are they still there or did you actually consciously throw some of them away? Some of them may have been worth throwing away!

As you grew older and went into school, suddenly there was a whole new sphere of influence upon you and your ethics and values. A whole bunch of other kids! The first five years of your life you were raised basically taking on all of the background your parents passed on to you. Now you're suddenly thrust into a whole different world; a world with dozens of other players. As you went on into High School you encountered hundreds or perhaps thousands of other players; people with ethics systems that were either similar to yours or completely different. It started to mold you and shape you and change you. It would be good to think back for a moment to consider how those things affected you.

Many people made many major decisions in their high school years that affected them the rest of their life. An example would be the decision to smoke pot or to get into the "drug culture" as it was called. This certainly led you down a different path. It exposed you to a different type of people than you would have had your decision been to say "no".

Your own personality certainly would affect you. For instance, what if you were the adventurous type as you got into high school or college? Were you willing to take on new challenges and excitement; go do something different and unique? These are things that continue to help mold you, build you. They expand on your experiences. Many of these type experiences directly affect your professional career. Even experiences that have nothing to do with your profession.

As you entered college you began to mold your decisions as to what profession you'd go into. That profession itself, especially as depicted by the text books and the instructors and professors that you had, began to mold you again. These people began to give you specific ways to apply (or perhaps not apply) certain principles that you came to the school with. You would be presented with certain professional issues having ethical implications. You would then have to react to them based on your personal value system.

Continuing with life's experiences, more factors enter into your ethical makeup. Whether you got married or not and whether you had children can greatly affect you as a person. Did that marriage last? Were you close to your children? These factors sometimes have a good effect and sometimes a bad effect on people. But it does change your approach.

If you had children and it was something you really enjoyed, you may have become a family man or woman. Then you started to look at things a little bit differently. You started realizing that there was some other purpose in your life than just to go out and make money. It may have caused you to slow down a little bit, which often times can be a good thing. This can contribute to a professional who's more methodical, more cautious, more aware of when a risk is being taken.

Once you got out of school and went into the profession another significant influence on the professional side of your ethics came along. It was the influence of your early mentor(s). The first surveying boss, party chief, or instrument person you worked with influenced you heavily. You went to the field and perhaps even lived with them a time or two out there. Those are people that had a direct impact on you. You started facing ethical decisions very quickly. You started facing decisions on what you were reporting you had done in the field. Distances were measured, angles were turned. Did you learn to be honest with those numbers? Did you learn correct procedures? Was the influence of those early mentors completely positive and proper? Only you can answer those questions. But all the while your values were being fine tuned and shaped by those you worked with.

Now, as you have become a professional land surveyor yourself you are a unique package, a product of thousands of influences. Some were significant and some not. But all of them have made you what you are. They established the professional decisions you will make. You cannot avoid it.

There are other factors that influence us. A big one is economics. We hate to admit it but economics are a real issue of life. They have heavily affected our professional ethics in the business world. We live in a society that coined the term "Situation Ethics" some thirty years ago. This is a line of reasoning that says that your ethics will change based on the situation. I can imagine a few circumstances where that would be true.

But this has been taken to an extreme; no matter what happens you take the course of action that is best for you personally. That may be your ethical value system. We live in a society made up of people who think that way. But is that really ethical? Is that choosing to do the right thing?

There are so many new issues now then there were twenty years ago that all professions have to face. We've been forced to look straight into the eye of some of our greatest

4

social weaknesses including racism and sexism. How you were raised has drastically affected your ability to cope with some of the changes that our society now demands that you make. This would include your ability to work in a multi-cultural organization. There are still many people who suffer grave things from their ethical background. They can't accept someone whom they know intellectually is just another human being like them. Their ethical background is so strong that it holds control over them. That skin color cannot be ignored through the eyes of their value system!

Nobody wants to be involved in sexual harassment. It is a demeaning to either male or female. But your personal ethics system has control over you! It affects how you look at members of the opposite sex or for that matter members of your own sex who have a different sexual orientation. You may not think these issues in any way relate to your professional ethics and conduct, but they most certainly do! 95% of the time ethics involve someone else. The reason you are treating someone in a certain way is an ethical issue.

Finally, another source of influence for your professional ethics is simply major events that occur in your career. Things such as who you work for, what kind of organization, whether it is a government, private sector, or utility. A major event that could effect you is to be called before the "board" to defend your survey. I know surveyors who came out of such challenges successfully, but very bitter! This "event" caused them to change their values somewhat. Being called into court can have a similar affect sometimes. I've seen people leave this profession because of such major events as these. Strange things can happen to us as we walk through this profession.

So let's answer the initial question in this chapter, "From where did your professional ethics come "? They came from your personal ethics. It is a complex discussion as to the sources of your personal ethics. I have briefly outlined some of the major influences in the average person's life. You as a professional land surveyor are a unique person. You are unique in the way you view the world. There is a unique set of factors that made you what you are.

Although we cannot completely understand and control those things, it is extremely important for us to occasionally reflect back on them. It helps us understand why we are the way we are when it comes to professional ethics. With this information as a background, the remaining chapters of this book are offered as thought-provokers. You must assess the issues discussed in your own way. But let us never forget that we are not working in a vacuum. We have a profession of which we are a part. Surely we owe some allegiance, responsibility, and courage to that profession. Your ethics reflect on all of us. So do mine. That should weigh heavily on our minds.

Chapter 2

TIMING

"Though I am not naturally honest, I am so sometimes by chance."
William Shakespeare

In 1975 I was working for a firm that was contracting from the U.S. Army Corps of Engineers. We were working on a highway and bridge project related to the Harry S. Truman Reservoir in southwestern Missouri. I was the party chief that essentially did all of the construction staking for this project. One of our tasks was to cross section borrow pits on a monthly basis. This was to provide the Corps with the material quantities used for determining progress payments.

One day the project manager for the contractor called me into his office. He was being kind and I sort of wondered what was wrong. People like that rarely give the surveyor the time of day. But he shot the bull with me for two or three minutes and then he finally got to the subject at hand.

The next day was to be our monthly cross sectioning of a borrow pit. This particular pit was approximately 25 acres in size. It was a major source used for "fill" out to a bridge across the main body of the Harry S. Truman Reservoir. The borrow pit was on private land and part of the deal was for the land owner to get himself a brand new 25 acre lake when all was done.

The project manager began to tell me how desperate the project needed some new equipment. In particular, they needed a new crane. Now cranes are rather expensive items, and I knew that. Also I knew that he was not going to ask me for a loan to buy that crane. I was working for five dollars an hour! He quickly got to the point; he wanted me to falsify my readings on the borrow pit. He basically wanted me to read everything off by two tenths of a foot. That way the two tenths of a foot over almost 25 acres at a dollar a cubic yard would provide some extra cash. It would help the company to go buy the crane they needed so badly.

6

He could see the concern and consternation written all over my face. At this point he said, "Well listen, next month when you cross section the borrow pit, you can read it exactly what it is and it'll just be short. It's just a matter of timing; we need the money now, not thirty days from now." I thought about it for a few moments as he continued to ramble on about how innocent all this was.

I'm sure there are many people that would not be bothered by this sort of action. After all, it all would have "come out in the wash". It would have been the exact same volumes from the start of the project to the end of the project. It was just a "matter of timing" of when they would get paid for that volume.

I very politely but strongly told him I simply could not be a part of this. He proceeded to insinuate that there would be some monetary reward for me if I went along with him. I reiterated that I was not interested in falsifying volumes.

Needless to say, my relationship with the project manager went down hill rather quickly. He never again spoke directly to me. In fact, within two months I was no longer working there. They no longer needed my "services".

What would you do? Did I do the right thing or was I being overly cautious? My greatest concern was that if I went ahead and did it, just thinking that it was a matter of timing, what would his next request have been? Would he have asked me to falsify volumes on a permanent basis? Or perhaps to cover up mistakes they may have made in location of certain items on the project? I didn't know but I knew one thing for certain; that was a "pit" I didn't want to jump into.

In some people's minds it's just a matter of timing. What do you think?

Chapter 3

JUST SA Y NO

"The first sign of maturity is the discovery that the volume knob also turns to the left."

"Smile" Zingers

I was running my own surveying business in Flagstaff, Arizona. I got a telephone call from the National Park Service. They were asking me to put a price in on some topography work up at Lake Powell. It was for the future Dangling Rope Marina.

Being an absolute lover of that part of the country I was thrilled and excited with that opportunity. I wanted so bad to work up there and I was even willing to do it at an extra low price. (This of course would fall into the chapter entitled "Low Bid", but for the time being I want to focus in on another aspect.) After negotiating a price I was awarded the job. But once that occurred I had to think really hard about the circumstance I was in. I was really quite busy. I was not able to take on that level of work, especially in a remote area. The travel time would put me completely out of touch with the office.

Finally, after two or three days of decision making, I had to call them back and simply turn down the work.

But there were many times in my career where I didn't do that, where I didn't say "NO" to the work. You probably know surveyors who operate the same way as I usually did. That is, you can never say no! You always take work; you greatly over-extend yourself at certain times.

It's easy to justify this. You can remember six months ago when you didn't have enough work. You were really suffering. So you figure you're going to "make hay while the sun shines." There's a real danger in the surveying profession in getting over-extended. It goes beyond the fact that you can't perform for the clients in a timely fashion.

It even goes beyond the fact that you would be just over-extending yourself in your personal time. It's more a matter of "can you truly do the quality, professional work needed for all of those people?"

I know surveyors who can never say "no." They could never turn down work no matter how big or small. No matter how much it was out of their area of expertise, they could never turn down work.

While it's nice to increase your market share of all the surveying work in your geographical area, it's also good to let the surveying profession serve all of it's clients with a high level of accuracy, quality and timeliness. It is an ethical question when one gets overextended for any reason. The real professional will recognize this situation and get out of it quickly.

When I called the Park Service back, they were very gracious. I thought they'd be ticked at me. But they were very understanding. In fact, I got another call from them two days later. They thanked me for being honest and not sticking them with an untimely contract. So I felt good about the experience....even though I missed out on surveying at Lake Powell!

Surveyors, we all need to learn that sometimes you just say "no!"

Chapter 4

TERRORISTS HOLD HOSTAGES

> "We must root out terrorism in this country."
> *President Clinton, after Oklahoma City bombing*

The latter half of the 20th century has seen many radical groups, both political and otherwise, who have used the tool of terrorism to be their means of accomplishing certain goals. It truly is a desperate and drastic action taken by people who basically feel that they don't have much to lose.

Terrorists generally are extremely disgruntled people who are willing to threaten the lives of innocent hostages, sometimes by the hundreds. They will hold the hostages lives in their hands and threaten to end their lives if their goals aren't met. In history we can even find where entire civilizations, entire nations have been held hostage. And we've seen plenty of proof where these people do not care about the lives of the hostages they hold.

The surveying profession has it's own version of terrorists. People who apparently don't have much to lose but everything to gain......if they can hold hostage entire groups of people, called neighborhoods. These are people whose professional ethics and conscience have been sold so that they can have personal profit.

I'm speaking of those who hold survey record information hostage. Many states have had the wisdom to enact mandatory plat recording laws. That begins to improve the situation. But a vast number of states have never instituted such laws. Even in the states where it has been instituted there's a vast backlog of unrecorded information. It simply predated the mandatory law.

Nothing is stopping surveyors from freely sharing their data between each other. Whether a law exists, or when it was enacted, is of no concern to the truly ethical surveyor. He or she will readily share their survey data. That's because they *know and understand* what surveying is all about. They recognize the *public service* provided by the surveying profession.

But there are many *unethical* surveyors who adamantly refuse to share data. For whatever reasons they may give, they willingly withhold important survey data that controls the boundary lines of many people. Entire neighborhoods are held hostage by surveyors who will not share that information. And others, the real terrorists, hold it for ransom. They demand money for their information.

The surveying profession exists as a licensed group of people for the express purpose of protecting the welfare and safety of the public. When you think about land surveying, how is it that the actions of the land surveyor would in any way effect the welfare of the public? It's easy to identify safety of the public with perhaps road construction and staking of buildings and various construction-related projects. Even mapping-related projects could fit that description.

But what about boundary work? That's where the welfare of the public really comes into play. The public licenses us so that their welfare can be protected. I submit that we are dealing with land, real estate, one of the most valuable things that any body can truly own. We must consider how to protect the welfare of the public with this valuable asset. It is the job of the land surveyor to identify boundaries.

Boundaries are limits of rights. They are the geographical location where rights begin or end. In order to know where rights exist on the ground, corner information is absolutely required. It is the public's right to know where corners of property are. The surveyor who does a project and does not share that information with the public has completely violated the purpose for his licensure.

There are many reasons why the land surveyor or engineering firm refuse to share survey information. One is, they are very concerned about liability. Obviously you would be more concerned about your liability if you had made a lot of poor surveys out there on the ground. But the fact is your liability has already been created if you set a monument. If there's something out there that you set you've already created a liability. Withholding the information about that monument in the ground or what monuments you may have found of others, actually *increases* your liability.

Another argument is, "this is my client's private information, he paid for it. It only goes to him." Certainly the client has a right to the information that he paid for. But let's remember again that we are licensed as surveyors to protect the public. That information (at least some of the information) must go to the public in a public record of some kind.

Another excuse that has been used to not share record information is that it's the surveyor's property and he doesn't have to give that up. The surveyor has produced a product, there's no question about that; the product is often a plat or a legal description. There are some good arguments for holding onto ownership of those certain items. But is the location of property corners private information that belongs to a surveyor? I don't think so!

There are hundreds of surveying and engineering firms across this country who have vast files and records in their possession. Much of it is produced by themselves and some of it bought from other firms that were going out of business. And many of those firms have been inundated by requests from other surveyors and even government agencies to get that information. Some surveying firms will hold hostage that information. They will ask for ridiculous fees or refuse absolutely to share that information under any circumstances. There is an ethical question involved here as to the whole purpose of surveying; the whole reason we're licensed; the whole reason we even exist.

I see no problem with surveyors charging a reasonable fee for the time and cost of researching their own records to provide that information. When you think about it, this could even be a new source of revenue for many a surveying or engineering firm. But the holding hostage of information where it's not released at all is unconscionable. It has been permanently kidnapped from the public. There is nothing that a surveyor does when he's dealing with boundaries that does not drastically affect other parties besides his client. That's why they license us; so everyone is protected......so we are tested......so we understand the laws and rules of equity, fairness, and proper procedure.

So how about you? What record information do you have now? What data will you have 25 years from now? Are you going to hold it hostage or are you going to be a really true professional?

The choice is yours!

Chapter 5

FLYING OVER KANSAS

"I don't think we're in Kansas anymore".
Dorothy to Toto, Wizard of Oz

In 1988 I was on a flight traveling from Denver, Colorado to Washington, D.C. I usually sit in an aisle seat simply because I like to have the ability to be able to get up whenever I want to. There was no one in the center seat but in the window seat was an attractive woman who was also traveling on business.

As often happens, we struck up a conversation. Those conversations usually involve swapping stories of where you're from, where you're going and what you do for a living. When she asked me what I did for a living I told her that I was a land surveyor and that I did some lecturing on the side in the area of the public land survey system (PLSS). This point intrigued her and she asked a few more questions. I explained that the PLSS was the rectangular grid of property boundaries that existed throughout the western two thirds of the United States.

At that point in our discussion we were flying over western Kansas on a beautiful, clear day. She looked out the window and noticed all the various squares and rectangles carved out on the earth below. These of course were roads running down section lines (or at least we assume they do!), fields that were in various stages of cultivation lying in forty and eighty acre configurations. After a moment she turned back to me and said, "You know, all those little squares down there, it's just so simple."

13

I didn't want to burst her bubble too much, but I mentioned to her that it looked simple from up here but the laws and the principles at work down there on the ground were really quite complex. That sufficed the question and the conversation turned to other subjects.

In my twenty-four years of working in the PLSS for both private and government interests, I have seen a wide array of survey practices and procedures in the PLSS. They have consisted of very good, high quality surveys. And they also consisted of very poor surveys that were blatantly illegal. I know from first hand experience that the vast majority of my fellow licensed professional surveyors do not know very much about the PLSS. In fact I felt they had a very simplistic view of that system. It seems to me that quite a few registered surveyors are just "flying over Kansas".

Perhaps my own personal experience would be of use as an example. My original survey license was in the state of Arizona and I received it in 1979. I had passed the twelve hour NCEE examination. I had not passed the four hour Arizona State Exam and so I repeated it and passed it on that second attempt. So now, here I was a licensed professional surveyor, I had my certificate on the wall, working for a firm in Flagstaff. About a year later I resigned from that firm over ethical issues and found myself running my own surveying business for about a year. I did all kinds of surveys all over the state of Arizona and I subdivided many a section. There were many, many circumstances that came up that I absolutely did not understand. In particular were strange lotting circumstances. I'm afraid I didn't have a very good handle on evidence or records research either. But, hey, I was a licensed land surveyor and with that plaque on the wall came a little bit of ego that really didn't belong.

In late 1980 I took a job with the United States Forest Service, also in northern Arizona. I worked for the Forest Service for about six months when my technical supervisor from Albuquerque recommended that I go to a four week long training course. That course is the well known Advanced Cadastral School that was at that time held in Tucson, Arizona.

Now I was willing to go down for some training. But four weeks seemed like tremendous overkill, especially four weeks of the PLSS. You see back then every time I worked in the PLSS I just took a short flight over Kansas and finished the job. Like I said, I was a licensed land surveyor.

By the end of that four weeks my jaw was on the floor. My eyes were now wide open. (And my hands were covering my rear end). I was absolutely shocked to find out how little I knew or understood of this system. I was quite humbled to be honest with you. But something happened to me at that session and in the weeks following. I suddenly matured as a land surveyor and made my greatest strides. It was at that point in time that I realized that if I didn't know very much about the PLSS (but had assumed I had), perhaps I didn't know as much as I needed to in a whole lot of other areas. It was then that I made a commitment to be involved in continuing education for myself. I was ignorant! It was six years later that I realized that vital information simply was not available to the average land surveyor and it had to be made available for our professions sake.

So I proceeded to do seminars, free of charge, for three years. All of these were on PLSS subjects. All of them with a deep sincere desire to get the word out. And the word was this; "we have a very weak apprenticeship system in the surveying profession". I personally believe that more than half of the registrants that we have (and that we continue to produce) do not have a basic fundamental grasp of many surveying processes and systems, the PLSS being one example.

The professional land surveyor who wants to operate in both a professional and ethical manner needs to ask him or herself two very important questions. They are:

- Through what process or processes did I learn the things that I
  know about surveying?

- Is there even the slightest possibility that I don't know every
  thing I should know about some or all of those subject areas?

If your answers to those questions cause you to perhaps step back just a little from the "I've got a survey license on my wall" point of view, then perhaps you should get off the plane over Kansas and take care of this situation. The truly ethical surveyor must know his limitations and must be constantly striving to improve them.

I'm happy to say, "I don't think I'm in Kansas anymore".

So where are you?

Chapter 6

THENCE

"Make sure also that you do not change those items which are already on record and if they do not make ties to any particular monuments or adjoining deeds, do not introduce new ones and claim them to be of record."

Gurdon Wattles

Surveyors should be familiar with the basic land description systems used to describe parcels of land. The intent of a legal description is to describe a parcel of land to the exclusion of all others. Legal descriptions have evolved over the years from very basic and simple description systems all the way to complex and formally designed systems, such as the Public Land Survey System.

It would be good to review those basic systems to prepare us for the discussion in this chapter. The basic description systems are:

Bounds: The Bounds description system was a simple process of describing who your adjoiners were at the time of writing. The land description would simply say that the parcel was bounded on a certain side by a certain person or a certain object. This system was used throughout Europe and the original American Colonies and Eastern Canada.

Metes and Bounds: This system is a more complex system than Bounds as it added "Metes" which are measurements of both direction and distance. Metes and Bounds is one of the popular systems used throughout the world today.

Portions of Another Parcel: This description system (Wattles called them "'LY De-scriptions") simply describes a parcel as being part of another record parcel. For instance: the Northerly one hundred feet of another parcel. This system has some

16

limitations and of course relies on there being another description system underlying its use.

Strip Descriptions: Surveyors are most familiar with Strip Descriptions as they are used for the describing of linear rights-of-way. For instance; roads, railroads or power lines. They generally require a consistent width and are described usually by their centerline.

Reference to a Record Document: This description system assumes that the parcel being conveyed is exactly the same as one that exists already in the record. It simply cites the location of the record and says "that same parcel as was conveyed there is being conveyed here".

The Public Lands Survey System (PLSS): Throughout the western 2/3 of the U.S. a survey description system was designed that was based on the rectangular grid known as the PLSS. The western two thirds of Canada had a similar system. (The system used in north and west Texas is similar but there are significant differences.)

Coordinates: A newer description system, not fully utilized throughout the world is that of reference to a coordinate system.

The land surveyor is supposed to be an absolute expert on legal descriptions. This comes from two aspects:

- Understanding the correct procedure of writing new legal descriptions

- The capability of interpreting older legal descriptions.

It is the duty of the land surveyor to explore and investigate thoroughly the "description history" used for a parcel of land and it's adjoiners. Legal descriptions are the heart and core of deeds. The deed may convey an interest in land and it contains the usual elements of a deed, such as consideration and acknowledgement. But the legal description itself tells specifically what area, what place, what location the interest conveyed is in.

The surveyor must be far more knowledgeable of this subject area than all the other users combined. Those other users include title companies, realtors, tax assessors, and lawyers. If ever there was a subject where the surveyor should be the absolute expert witness, it's in the area of legal descriptions.

The surveyor should also be fully versed in the processes and legal principles involved with each of these land description systems. They differ significantly. There is considerable case law regarding the interpretation of various elements of these different land description systems. In fact, the most complex and complicated thing that a land surveyor ever deals with is in the of arena legal land descriptions.

Why is it then that we spend so little time preparing ourselves to know about these description systems? Why is it that we often turn over legal description writing and interpretation to people in the lower levels of our profession? Why is it that we have so little formal training in the complexities of legal description interpretation and construction?

I've said several times in seminars and I still mean it; the only requirement to write a legal description is the ability to spell the word, "thence". I say this for two reasons. First of all, judging from the vast majority of legal descriptions that I've had to work with, the persons writing them had little or no idea of what the legal principles involved were. The second reason is even more significant and that is the purpose of this chapter.

Of the seven description systems listed above which one do you believe is most often utilized? In my experience in many different states it is undoubtedly the metes and bounds description. Of course, it is within the metes and bounds system that one would encounter the word "Thence". Thence is the connecting term between calls in a legal description. It indicates that you have completed the previous course and now you've started another course along the exterior of the parcel of land being described. It is that intense desire to say "thence a bearing and distance"; "thence another bearing and distance"; that has driven our profession far away from the intent of the metes and bounds legal description. It has amazed me how many times we have resorted to a metes and bounds description when one of the other descriptions systems would have been far better. Often a parcel of land has been historically described by another description system and we choose to go to metes and bounds instead.

The classic example of this is the converting of a PLSS description into a metes and bounds description. Anyone who does this is broadcasting their ignorance of legal description systems and the legal principles involved. One cannot rewrite a PLSS description into a metes and bounds description without great care and concern. You must be sure the legal principles needed are still in that description to protect the intent. The surveyor who thinks that the southwest quarter of the southwest quarter of a section is exactly the same as "beginning at the section corner thence North 1320 thence East 1320 thence West 1320 thence South 1320, to the point of beginning" is truly mistaken. These are not the same parcels of land. The legal principles behind those two description systems are quite different. One could assume that they are exactly the same parcel of land. That's a major assumption and there are many circumstances where that would not be true. There are also many court cases that say that is not true.

It seems there is a great comfort of writing a well structured metes and bounds description that has "Thence" and "Bearings" and "Distances". But we should not be lulled into a false sense of security with that. A metes and bounds description serves a great purpose. But in many cases it has been abused and misused. It simply is misapplied to parcels of land that it should have never been used upon.

A tremendous amount of time and effort is wasted in writing metes and bounds descriptions when the existing record description is perfectly adequate. The record descriptions may be just as legal and binding and clear. For instance, suppose you've got a subdivision plat that's filed with the appropriate public authority. A deed refers to lot number 3 of block 2. Why would you need to rewrite that description as a metes and bounds and go all the way around the outside of that parcel? It is perfectly described by reference to the record document. Taking this action again shows an ignorance of the description systems and their purposes.

But our comfort level is increased when we get to put in bearings to the second and distances to the hundredth (or even thousandth of a foot) and areas of land to the ten thousandth of an acre. Let's not be fooled. This profession should be the experts on legal land descriptions, but we're not.

It's time for each of us to analyze very carefully our abilities in the legal description arena. There are many good text books on the subject. Yet I dare say, from my own personal experience of reviewing thousands of legal descriptions, very few people in our profession have ever read one of those text books all the way through. I would strongly suggest that people would look at Gurdon Wattles' book entitled "Writing Legal Descriptions". That one book alone has more information than the average surveyor will need. We don't have that information working on a daily basis in our profession.

So what about you? Do you know how to spell "thence"? Is that the only requirement to be met? What makes you and I an expert on legal descriptions? It should not be based on our ability to convert anything into metes and bounds with the addition of a bearing and a distance. Rather, our being experts on legal descriptions should be based solely on our complete and thorough understanding of the legal principles behind each of the land description systems.

We've been "thenced to death". Let's become the experts we claim to be. It is an ethical question. If we hold ourselves out as the experts, we better *be* the experts.

19

Chapter 7

DOUBLE MONUMENTATION

"No...resurvey or retracement shall be so executed as to impair the bona fide rights or claims of any....owner of lands affected by such resurvey or retracement."

The Act of March 3, 1909 (43 U.S.C. 772)

The world of land survey "users" has some interesting perceptions of the profession. Most people believe once you hire a surveyor, your boundaries are fixed and certain. We as surveyors should recognize that this is often not the case. We can report the facts. Sometimes the facts are quite clear and a boundary is very safe and secure. Other times, the appearance of possible unwritten rights, both for or against a client, can lead the surveyor to a less certain conclusion.

Such problems should be thoroughly represented on the plat and also explained to the client. But occasionally surveyors themselves create or perpetuate illogical conclusions or gray areas which need not exist at all. These actions cause a great deal of heartburn to the client, and well they should. I'm speaking of the practice of double monumentation.

Double monumentation occurs in two basic ways. First, the surveyor disagrees with a found monument and so decides to "set the correct" corner. Second, a surveyor notes on the plat that a found monument is a certain distance or move from the "true point", but does not set another monument. The purpose of this article is to discuss these two common practices.

When should a second monument be set? One of the most damaging things the surveying profession can do to itself is allow the setting of double monuments. As many of you have seen, this can sometimes be triple or more monuments all claiming to be the same corner point. This sends a terrible message to the public. We look like a bunch of prima dona's all trying to out-brag each other about our accuracies and precisions. I often expect to see those surveyors swinging on vines and beating their chests. They try to tell the world "My measurement is better than yours." Some even brag on their plats how they adjusted their traverses or how well they closed.

While on the surface this all may seem professional, it is actually about as far from professional as you can get.

The heart of this issue is simple; when should you accept an existing monument and when should you reject it? This question touches the very basic reasons we have surveyors and why we license them. It is my observation that most of the time a second (or third or...) monument is set is for entirely the wrong reason. If a deed description calls for a monument, and you find that monument, some very basic principles of land law say you should accept it. This is true even if it is not at the exact called for distance or bearing. The call for a monument is the most powerful call in any description. But some "measurers" are apparently not aware of this concept. They will set their own "superior" monument a half foot away and create confusion where none need exist.

Often, an uncalled-for monument is found at a point where a deed has taken the surveyor. We must not automatically reject such a monument simply because it does not fit our precise measurements. An element of common sense must enter in when dealing with the uncalled-for monument. Does the position monumented mark the corner point within reason?

For example, consider a description you are retracing that calls for "thence North 150 feet" to a property corner. Your survey locates an iron pipe at this approximate point. But you calculate the pipe to be about 6 minutes off of bearing and .30 feet too far. Should you set another? You must ask some questions about this uncalled for monument.

- Is there a record of this monument?
- Where did it come from, who set it?
- What deed (or deeds) were being used to establish this position?
- Is there a conflict with those deeds and your deed?
- How long has it been here?
- Was it set with reasonable accuracy given all the circumstances?
- Will it better serve the public by setting an additional monument?
- Who and what has relied upon this position and for how long?
- Has acquiescence taken place anyway?

I have seen five monuments within a two square foot area, all purporting to be the same corner point. They were based on a legal description 75 years old in heavy mountainous and wooded terrain. Is there no common sense left in our profession? Are we being reduced down to nothing but a bunch of precision-hogs?

21

In the previous example, the bearing error is meaningless unless you have been very prudent to know the basis of bearings you are retracing and the factors that may have influenced the "precision" of the record bearings. Let's face it, 6 minutes in 150 feet is only 0.26 feet on the ground. Is it not within reason this monument is a reasonable attempt to mark the corner point? Was the distance so unreasonable? Do you really think your traverse, after being adjusted, is all that significantly different? A pipe in the ground for 25 years should carry a lot of weight with the present-day surveyor if it reasonably marks the originally intended corner point.

But there are situations where the surveyor must set another monument. When the found "uncalled-for" monument is not within reason, or appears to have been set incorrectly (notice I did not say inaccurately), then a second monument may need to be set. Your plat should clearly state the fact that you reject the found corner, and explain *why*!

The multiple monument syndrome is usually not a result of careful consideration, research, and professionally-based common sense. Rather, it is the result of measurement, adjustment, or technique disagreements. In short, a second monument is only needed when there is a *clear* reason why the found point should be rejected. It should never be based on reasons of accuracy, measurement, or precision alone. When you consider the principles of why a monument is so "sanctified" in land boundary law, you will understand why precision was never a factor in land surveying. Harmony with record angles, distances, and areas is a major test, but these must be realistic and within reason.

Some will misread this to say I am not in favor of good survey practices or precision. This is simply not true. The real issues in property surveying are legal, not mathematical. When we cross the line and worry more about precision than "right", we undo the purpose of our profession and literally curse the public whom we are supposedly protecting.

Some very good reading on this subject is sections 4.22 and 5.16 of *Boundary Control and Legal Principles*, 2nd edition, by Curtis Brown. He discusses both the uncalled-for monument in a metes and bounds and in a simultaneous conveyance situation.

So what about the un-set "true point"? The second practice, mentioned earlier, is the plat that states they "found a rebar", but the true point is ".04 North, .07 West of the rebar". Really? Who are they trying to fool? I know of firms that do this all the time, and yet they have never heard of a prism offset. They never adjust their tribrachs and they have no concept of positional tolerances. Their traverse closes 1:20000 over a four mile length, but there is no way their work is even remotely accurate enough to make this kind of judgment.

These licensed computation artists (COGO slaves) are a further detriment to an honorable profession. This practice defies all logic.The public cries out, "where is the corner"? Is it the rebar, or the theoretical point? What service does this do the client, the adjoiners, or the public in general? None.

At what point in time does the monument finally take its rightful sanctity and become the corner? Should we wait until every surveyor in town agrees to it within .01 foot? Hold on, GPS may help us bring it down even more. When will it end? Some surveyors cannot seem to make a commitment. Is it the corner or not? That is what they hired you to do.

When the circumstances discussed earlier are present, there may be a good reason to set a second monument, but this should be the exception rather than the rule. The Test of Professionalism: While various groups and organizations for surveyors continue to roam the country preaching the paths to professionalism, there has always been only one true test; the quality of the work you perform. The setting of double monuments is almost always a sure sign some amateur has been there before you. Perhaps they left a real record of "why" on a plat. But usually you simply find a game of multiple choice at the corner point. Choosing the oldest monument is not always the solution.

These games are creating more and more disputes, controversies, and ill-will toward the surveying profession. I urge all who read this to consider what they are doing. Are you really practicing surveying? Or just a mathematical shadow of the profession?

Chapter 8

THE FIRST SURVEYOR

"Cursed be he that removes his neighbors landmark, and all the people shall say amen."

Deuteronomy 27:27

Have you ever wondered who was the very first surveyor? It might be a surprise to you. This person wrote an amazing book which in part documents this surveyor's purposes and methodology. That First Surveyor was God, and His book is the Bible.

You may have never realized this before, but the Bible has many references and direct commands about the subjects of land ownership, boundary law, business practices, and legal descriptions. Whether you credit what has been called the "Book of books" with divine authorship or not, you cannot deny the incredible insight on issues of interest to our profession.

It has been said the first subdivision was created in Genesis 1, when God "divided" the heaven from the earth. I've also heard that the marked stones that Moses brought down from the mountain were actually section corners. These are stretching things a little, but let's look at some real references that may give you a little more feeling of "heritage" for your profession.

The book of Genesis did start in motion some basic paths of the land development (and thereafter surveying) business. There was a specific command to "be fruitful and multiply....and to subdue and have dominion over the Earth." While it may seem some have taken advantage of this command to the detriment of the planet, it does bring out a desire on the part of the Almighty to give man control over his own physical destiny. Taking this command in balance with the rest of the principles of this book, it is clear the intent was an orderly, proper, and well planned development of the Earth. The fact that in many places it has not turned out that way does not nullify the original intent.

The story is picked up in the book of Exodus, where Israel, God's chosen nation (a model nation if you will), started out as a nation of slaves escaping from Egypt. As they arrived in the "promised land" (some 40 years later) an interesting event took place. God, through Moses, provided legal descriptions to the 12 tribes so the land could be divided amongst them. This can be found in Numbers, chapter 34. Note the particular use of "natural monuments" in the legal. In verses 17 through 29 God specifically names the survey and mapping team which would lay out all this work.

Once the nation of Israel was established, God gave them several principles to help maintain an orderly system of land tenure. First, the "land sabbath" was created (Leviticus 25), a rest from agricultural production for the land every seventh year. Not a bad idea for this "modern" 20th century!

Further, the year of "jubilee" was established. This was a law that said, every 50 years, all land must be returned to it's original family of ownership. This prevented long term land speculation and over-development. Sounds nice doesn't it? This would certainly prevent a scoundrel in one's genealogy from permanently giving away the family heritage.

In Deuteronomy (which literally means the "second giving of the law") chapter 19, verse 14 (19:14) a discussion is made on the importance of preserving ancient survey monuments. In 27:27, the law says "Cursed be he that removes his neighbors landmark."

The book of Proverbs cites a similar surveying-specific law. Notice chapter 22:28, where we find the principles of a monument being the most senior call. Truly the monument is given a "sanctified" position in the list of calls.

Somewhat in jest, I point out the fact that God did not make a law about crimping your neighbors chain, or changing the declination of his compass. It was not a measurement issue that He was concerned in protecting.... it was the location on the ground that counted. Have you ever wondered where that common law principle came from? Now you know. Even today, the basic rules of the "seniority of calls" say a monument will override a bearing or a distance.

Several pertinent business practices are mentioned in the Bible. One which directly relates to the surveying profession is found in Deuteronomy 25:13-16. Here are commands about "differing weights and measures" in your business. Having our measurements honest and "realistic" is made more than a topic for a "situation ethics" discussion. It was the Law!

To purposely withhold critical information is condemned in Leviticus 5:1. And any monies gained through fraud, deceit, or incompetence were to be returned in full, plus a 20% fine.

Even professional responsibility (or liability) is mentioned in Exodus 21:28. The passage covers a death caused by an ox. A warning is given that if the ox was known to be dangerous but was not dealt with, there is a serious liability created by the owners thereof. The analogy to engineering and surveying is obvious.

A different type reference to our profession in the Bible is the mention of some of our tools in prophetic warnings. A plumb bob is used in a prophetic analogy in chapter 7 of Amos. Here God makes reference to the plumb line and says His Law (the ten commandments) will be used in a similar fashion to measure the uprightness of the people.

Also a prophecy in Zechariah 2:1-3 uses a measuring stick to compare the peoples' obedience to the Law and to God's other principles of living. A similar passage is found in Revelation 11, referring to a "rod" to measure the people in the temple.

In the book of Job, God challenges Job's self-righteousness by asking him "Where were you when I laid the foundations of the Earth?.....Who has laid the measures thereof?...Who has stretched the line upon it?.....Who laid the corner stone?"

Finally, when Christ was on this Earth, the profession he learned from Joseph was that of carpenter. We should not lose sight of the fact that a carpenter back then was much more than one now. Working with wood was only the beginning. He was also trained in masonry, plumbing, and the sciences of engineering, architecture, and surveying.

I suppose the one statement in the entire Book which summarizes all the business, surveying, and ethical issues presented is Matthew 7:12. Here we find what has been called the "golden rule". "Therefore what things would you have men do to you, do you even so to them....". If our profession (or world for that matter) would heed only these few words, things would be different!

These numerous references convince me the first surveyor was God. He says He laid out the universe, the planet, the land divisions between tribes (nations), and established principles of land ownership and boundary determination. He provides business principles which span the ages with their simplicity and truth. Further, He draws analogies utilizing surveying tools to show how to live.

As I look around me, at all the amazing structure, pattern, and organization I see in the creation, I cannot deny that a surveyor has been at work long before me. And I pause for a moment and reflect on the eternal heritage my profession has behind it. I'm proud to be a professional land surveyor. I do my job well. But I'll never forget to pay homage to the *First Surveyor*.

Chapter 9

ETHICS IN THE COURTROOM

"In modern trials, while the lay witness testifies to what he saw, or, more correctly put, to what he believes he saw, the expert witness, with cold, calculable facts, figures, and photographs, answers objectively to the question."
Melvin Belli

The last place a surveyor ever wants to find him or herself is in court. Whether it is over a business deal or your own survey, the legal arena is no place for any nice person to be. I know; I've been in court a few times in a wide variety of types of cases. It has never been pleasant.

Some in our profession think it is the ultimate opportunity to be important; to be an "expert witness". While this duty is an important part of our professional service, it is never where we should want to see things get resolved. My personal philosophy has always been to try to avoid the courtroom at all costs. You cannot avoid court completely, but there are some things the surveyor can do to minimize those experiences.

Perhaps the single most important means of avoiding court is *communication*. I mean good thorough open communication. It has always amazed me to see how often issues go before a judge or jury where the facts were never clearly stated to the parties to begin with. I've seen folks file lawsuits over complete misunderstandings; their surveyor had not really explained things properly. I believe good communications have kept me out of court dozens of times. I made sure the "other side" of a dispute had every opportunity to see that I was honest, open, and fair.

I remember a case where I was sitting in a room with my client, and his neighbor was there, with a lawyer and another surveyor. The other surveyor was helping them attack a survey I had made which had shown a major difference in the location of a section line. That surveyor had simply compared numbers on plats and had concluded my survey had "moved a section line".

The attorney was doing his best to serve his client. He had been given so little information from the surveyor, I was embarrassed for our profession. The attorney knew almost nothing about land boundary law, and apparently the surveyor did not either. He simply made some allegations, and here we were discussing the issue to decide if a legal battle was necessary.

I decided right there to "educate" that attorney; something his surveyor had failed to do. I methodically went through the survey process, the legal issues at hand, and the evidence that had been found on the ground. He listened intently. I convinced that attorney that I was a real expert in the subject area, and that I was fully prepared to discuss any aspect.

The other surveyor was getting agitated. He kept making wild accusations, threats, and slanderous innuendoes. He was so busy justifying his survey, his existence, and his importance, that he did not notice the change in his attorney's demeanor. I really believe the attorney started to give me more credibility than his own surveyor. Within an hour, the attorney finally turned to his own surveyor and said, "why don't you just shut up!". Good communication skills mixed with patience will often win out. I believe this to be an obligation the surveyor has to his client and his profession; learn to communicate.

But what about the times you don't succeed in avoiding the courtroom? Perhaps you have received a subpoena for a case in which you were never directly involved? I recently testified (for three long hours) in a criminal case. I never thought I'd be in court over that survey, but there I was. Three hours on the stand, and the survey wasn't even the central issue to the case. But I took one big beating from the defense attorney, even though my survey had absolutely no flaws! Remember; when you go into court, you're not surveying anymore.

The legal system is a great mystery to most of us. Over the years I've come to understand it more fully, but it still irks me how it works. I hate going to court. It is not my purpose in this chapter to give you all the do's and don'ts for preparing for and appearing in court. There are some excellent references available on that, and I would strongly suggest a textbook by John Briscoe entitled "Surveying the Courtroom".[1]

But there are some ethical issues to be discussed here. They are simple, but powerful. Any good attorney will tell you your ethical responsibility in the courtroom is simply to "always tell the truth". That is easier said than done when a defense attorney is firing multi-faceted leading questions at you. You rarely have time to explain what you think you might have said!

Always tell the truth. That ethical principle must be obeyed. If it is not, you risk a serious breach of your testimony, if not your credibility as a professional. Let's look at a few areas where this principle must be present in the arena of legal battles.

---

[1] "Surveying the Courtroom", John Briscoe, Landmark Enterprises, Rancho Cordova, CA 1984.

- Tell your attorney the truth about your qualifications. Do not overstate them or try to compete with the expert witness on the other side.

- Tell your attorney the truth about your past. I recently saw a hydrologist get ripped to shreds by an attorney in a federal case. He had been fired by the federal agency he was testifying against. He was a fool to think they would not use this in the battle to discredit his testimony. The hydrologist's attorney was caught completely off guard. There was nothing he could do to defend that witness. And I'm not kidding, there was blood and guts all over that courtroom that day from his "expert" testimony.

- When confronted with issues such as mentioned above, admit it immediately on the witness stand. The hydrologist denied he was fired, and insisted he was telling the truth. When the opposing attorney brought out the subpoenaed personnel files, things went to hell real fast. They even entered the hydrologist's own sworn affidavit into the record, where he admitted to 15 counts of theft from the agency. Needless to say, his testimony was more damaging than beneficial to the client.

- Tell the truth on the witness stand. Don't exaggerate, don't understate, just tell the truth. One of the most important elements of telling the truth on the stand is to say "I don't know". I guess we feel as expert witnesses that we have to have an answer to everything. But if you don't know for sure don't say it. (I recently was asked on the stand if I had done a solar observation on the survey in question. I knew we had but for some reason it did not compute correctly, so we used an assumed basis of bearings instead. I answered "no" to his question, but thought later of how that could have backfired on me had the attorney been more aware of my fieldnotes he had also subpoenaed. I was lucky that time.)

- In your dealings with the "other side" of a case, whether in communication, deposition, or other elements of the discovery process, be clear, concise, and honest. Things you slightly exaggerated in an affidavit or deposition will come back to get you later on the stand.

Finally, whatever you do, don't let your ego get in the way of common sense. I have a few times, and it always played against my testimony. The expert witness is supposed to be unwavered by who his client is, or what "side" he is on. Our testimony should be at the same professional level and content, no matter what the subject, who the client, how much money is at stake, or who the other surveyor is. My ego would flare up and try to beat that other lying surveyor into the ground. But juries and judges usually see right through that. So if your survey is under attack, you must calmly and humbly defend it, but never argue it.

There are professional ethics in the courtroom, and the most important is to tell the truth. That's good advice for us at any time in any circumstance.

Chapter 10

FOR CHARLIE

"Iron sharpens iron; so a man sharpens the countenance of his friend."

Proverbs 27:17

I met Charlie in 1979. We discussed a project that I was going to do for him. He was an excellent land surveyor, a very well known land surveyor in the southwestern United States. We discussed a number of items in the work that was to done for him. It kind of excited me that he was quite impressed with the proposal that I had put together. I was honored that he was willing to put me on to a rather large cadastral project in Arizona for the agency that he worked for.

Charlie and I spent the night in a small town in Arizona. We went to the bar (as surveyors often do) and I noticed that Charlie quickly drank himself right out of his mind. For another ten years I knew Charlie, spent time with him, had many meetings with him. He was always the life of the party; but always a heavy drinker. I remember when he finally retired from the federal agency for which he had worked for, some 30 years. He moved to a small town where he was going to live out retirement with his wife and a relatively young son. But I also knew that a vacuum was going to be created in Charlie's life because of the loss of the job. It would have a serious impact on him. Charlie was a heavy drinker, even in the good times. What would become of Charlie now?

I lost touch with him over the last couple of years of his life. He lived some two hundred miles away from me so we didn't have as much contact. But I knew that Charlie continued to have problems.

It was in 1989 when I was planned to go see Charlie at a surveyors meeting. I looked forward to visiting with him and asking him how things were going. But three days before that meeting Charlie put a gun to his head and killed himself. The shock and the dismay was overwhelming for many of us who had known him for so many years.

The profession had lost a solid member. He was a very insightful land surveyor. In the ensuing weeks it became clear that Charlie's drinking problem had gone far beyond where it had been before. He was even to the point of being extremely abusive of that child that he loved so much.

His retirement; their retirement, had now become a nightmare. Apparently his wife and child had left him just a few weeks before this meeting. Charlie was so depressed and despondent and so embarrassed to have to face all of us, all his old friends. It would inevitably come out that his life was absolutely meaningless.

I don't write this chapter to be critical of Charlie. All of us have our crosses to bear and Charlie's life was one of constant repeated tragedy. His first wife had died of cancer. His oldest daughter had been murdered at college and his second wife divorced him. Yet still he was a fine professional.

I'm not personally opposed to drinking but I throw this thought out: Surveyors (along with many other professions) seem to pride themselves in volume of alcohol consumed. They have this cultural paradigm that they're big heavy drinkers, they can really drink hard. All of that is just vanity and ego when you boil it down. I wonder how much of that kind of professional cultural influence affected Charlie? I wonder how that drove him on? I wondered how that drove many of us on to do things that we normally wouldn't have done that were detrimental to us and the profession.

I don't want Charlie to have died in vain. He was a good man. I loved the man. But you who are reading this must ask: Is alcohol really controlling my life? Or drugs? I offer this thought for you to consider; The ethics for you as a person, for those who love you, and for the rest of us in your profession are at stake. Is it really worth it? I would hate to see anymore people have to end their life like Charlie did. This is for Charlie. A good surveyor. A professional. But alcohol stole him from us. What about you?

Chapter 11

PRICE FIXING

"The Northwest Chapter of ASPS is alive and well. The chapter is looking forward to less exciting activities in the future."

Steve Hankins, President of ASPS (Arkansas)

Recently our fellow professionals in the state of Arkansas learned a very hard and bitter lesson about price fixing. Many "Monday morning quarterbacks" have jumped up saying "Yeah, they shouldn't have done anything like that. They violated all these laws and rules and they should be punished."

I cannot condone actual price fixing. The anti-trust laws of the United States (primarily the Sherman Act) are there to try to provide some equity in competition and fairness in price from groups of service or product providers. Price fixing involves the purposeful setting of prices by a group of providers. It involves meetings and agreements to lock prices in a way that provides no choice or fairness to the consumer. So let's settle it right here......price fixing is illegal and unethical.

It is important to realize however, that the Arkansas case had nothing to do with real price fixing. An overly-zealous attorney in the Justice Department saw a great opportunity to "nail" a professional society. This was something rarely ever accomplished. And unfortunately, the surveyors involved settled out of court in a way that admitted they were in violation of the anti-trust laws. But they in fact were not in violation.

I understand what they were trying to do. The surveyors in the northwestern part of the state recognized, as do most of us throughout the rest of country, that surveying fees are ridiculously low. They knew that other entities had strong influence on their pricing. In particular, the title companies always showed the estimated cost of a survey on their "Estimate of Closing Costs" form.

Surveys, improvement location certificates, or mortgage surveys (or your equivalent) are being done at rock bottom prices. The surveyors of the Northwestern chapter of ASPS were trying to get a reasonable base fee established for that kind of work for estimation purposes for the title companies. A base fee already existed in the area. The title companies already had a base fee estimate of $100 they put on everyone's estimate of closing costs form. These surveyors simply wanted the title companies to show a higher estimate. If anti-trust was at work, it had always been there.

If the Justice Department was really concerned about price fixing they should have considered the gasoline industry. Have you ever seen prices differ by more than a penny a gallon between two gas stations across the street from one another? One goes up, the other one goes up. If one goes down the other one goes down as well. Our friends in Arkansas were beat up pretty good for a very minor infraction. They were trying to solve a real problem. So obviously price fixing is off-limits in the way that would violate those laws.

The real problem is not the price. The root cause of the problem is that there are people doing sub-standard work. Let's face it, you can't do a really good survey for a hundred bucks. The only way you could is if you just happened to have been in there a few weeks ago and know everything about it. There's a lot of places in this country where entire sections are supposedly being sub-divided for 500 dollars. Improvement locations certificates or mortgage surveys are being done for 50 dollars.

I have never ceased to have been amazed at the value system that society has placed (and that we have allowed them to place) on a survey. Imagine if you will a parcel of undeveloped land in a rural area, perhaps 40 acres. Imagine that the owners of that 40 acres have decided to sell that land. So they list it with a realtor, a licensed professional. A few weeks go by and the realtor sells that piece of land. Let's just say that the land is worth about a thousand dollars an acre. So he sells this land for forty thousand dollars.

The standard realtors commission in the Untied States for undeveloped land is 10%. (If this isn't the epitome of price-fixing, I don't know what is!) The realtor made $4,000. Now the realtor did have some expenses. He probably spent $50 in newspaper ads. He spent another $50 in gasoline driving people out to the land in his Cadillac to show them the piece of land. And he probably spent another ten dollars on getting his shoes cleaned and repolished because of the mud that he stepped in when he went to show the property. So he hasn't got much invested but he has made $4,000 and nobody complains.

They decide to get a survey done. They hire a land surveyor to come in and break out this 40 acres. The surveyor gives them a fair and equitable price. He quotes them and estimates that it will cost them $1500 to do that survey. Then listen to the outcry and scream! Something is wrong and we've allowed it to get there. The surveyor will be lucky to make a dime of profit on that $1500 fee for that survey, and yet our society absolutely goes bananas at that kind of a price for a survey.

What's even more frightening is, that guy is going to panic at the $1500 quote and he's going to play "yellow page roulette" and hire another firm who will do it for a thousand. Now you can claim all you want that, "That's the great American way," but I'm here to tell you, that is not how any profession should operate.

No one is completely sure who blew the whistle on the big bad surveyors of north-western Arkansas. But the odds are a million to one that it was either a title company or a realtor, or both. Think of it....either a price-fixed realtor reported them, or else a title company who should have no care on the closing costs did. Either way, surveyors are not running their own profession!

So how can we get rid of the low ball price? How can we solve the problem that our friends in Arkansas were trying to deal with? The answer is a difficult and time consuming process, but it must be done. Some suggestions follow:

Get rid of the bad practitioner: We have many surveyors out there that are doing very poor work. They don't have to charge all that much to do a good job because they don't do a good job. We've got to identify who those people are and quite frankly get rid of them. They need to be expelled from our profession. The sooner the better. That's not easy and it's going to take a lot of guts. (See the chapter entitled "The Low Bid")

Require mandatory continuing education: We have a chapter in the book on that subject. But in this regard it will help educate the bad practitioners as to the ethical and legal requirements of their profession. Unfortunately a large portion of our profession does not have any idea of what the requirements are.

Establish a mandatory plat recording law: This will solve a myriad of problems. It will also stop the poor practitioner. One of the arguments I've heard against mandatory plat recording is, "Oh no, it's going to drive the cost of a survey up, people can't afford what we're doing now." And yet that same person will be complaining the next day about how poorly surveyors are paid.

Price fixing is illegal and I don't encourage any one to do it. But establishing professional standards, minimum levels of operation and decent work, and eliminating the bad practitioner, is the only way that we can see our prices rise. We provide a tremendous service far more important, far more lasting, far more permanent than the realtor, the appraiser, or the title company will ever do. And it's time we started charging the right price. You don't have to meet with your fellow surveyors in a back room and set the price. All you need to do is as an individual say enough is enough, we're going to charge what it's worth and we're going to bring the rest of the profession up with us. We will not drag it all down to the low bid.

Let's eliminate the real problem, permanently. And as for our fellow surveyors in Arkansas, I salute you, I honor you, and I support you.

Chapter 12
CADASTRAL WARS

"Our task is not to fix the blame for the past, but to fix the course for the future."

President John F. Kennedy

Not so long ago...... in a galaxy not so far away, there was a battle between two federal government agencies who perform land surveying operations. I'm speaking of the Bureau of Land Management and the Forest Service. These two agencies have long been, in different ways, the leaders to the surveying profession. The BLM obviously is the keeper of the Public Land Survey System (PLSS). They are the modern-day version of the General Land Office with whom the PLSS started. They continue to hold what is called "federal survey authority". This is meant to represent the only authority granted by the federal government to survey. It included the power to create the public land survey system.

The Forest Service on the other hand, is an agency that has existed for about 100 years. In the 1950's they began a land surveying program which they called "land line location". The Forest Service program was to go beyond what the BLM and GLO had done. It was to further identify the private inholdings within the National Forest System. This went beyond the original scope of the GLO and the BLM surveys. It also involved the marking and the posting of those boundaries so that any non-surveyor could readily identify a boundary in the field.

A bitter rivalry developed between these two agencies. Much of it was centered on who had authority to do what. The BLM having federal survey authority felt they were the sole surveyors of land in the United States that involved public domain. The Forest Service, on the other hand, acknowledged that they did not have federal survey authority. So they hired or contracted with state licensed surveyors. Either way, both agencies have some kind of survey authority whether some individuals in either agency want to admit it or not. The battles between the agencies became quite vocal and quite public. It was embarrassing for many of us in these two agencies.

Both the Forest Service and BLM have some wonderful, highly skilled professional people. The vast majority of these people get along with one another very well. However, at times, certain individuals (including the leaders of those agencies and surveying divisions within those agencies) have not got along well and have made extravagant efforts to stir the pot. There have been insults made publicly in meetings and constant putting down of one another. It was a sort of one-upmanship.

About 1990, the BLM embarked on another federal survey authority "jihad" to prove their sovereignty over the universe. Both agencies attempted to find poor surveys the others had done. They pointed them out to others, and even brought them to public meetings by airing the other's dirty laundry.

Some common sense has taken over both agencies in the last few years, and that is a welcome relief. Each agency should be acting like the co-sponsors and co-leaders of the land surveying profession especially as it pertains to the public land survey system. They should not be in one giant battle with one another, wasting tax payers dollars researching various things and trying to prove who's best.

This subject is near and dear to my heart. I have many friends in both agencies. I also have deep respect for both agencies. This is a cry that we set aside the cadastral wars. We must start cooperating even more. We must find ways to truly assume our leadership roles.

Let's shoot Darth Vader, Jabba the Hut, and the Imperial Storm Troopers. Let's get on with the governments' survey needs. Let's act like the responsible leaders of a great profession. The taxpayers demand it. And so does our own profession.

Chapter 13

BEYOND YOUR BOUNDARIES

"Things which matter most must never be at the mercy of things that matter least."

Goethe

Lin Livermore (a former BLM employee) in a speech he once gave, said, "On your way down to this meeting, did you think of how many boundaries you had to cross?" That's an amazing statement because it really causes you to think about how intricately involved this society is with our profession. On your drive to work, you cross hundreds (if not thousands) of boundaries. Property boundaries, easement boundaries, and those of other types of rights and governmental authorities. When I think back to the forefathers of surveyors in this country (i.e.. George Washington, Thomas Jefferson, Abraham Lincoln etc..) I realize how fundamental surveying is to the fabric of a civilized country.

Surveyors are very aware of boundaries. We are very attuned to rights and the limitations of those rights and the limitations of opportunities that boundaries can create. Just think of the guy who builds his garage across the property line by mistake. If he gets caught, his opportunities have been greatly limited. (I suppose if he gets some kind of adverse possession out of it, he actually increased his opportunities!) We are very aware of those sort of things.

I wonder if we surveyors suffer from an inability to see "beyond our boundaries." I mean our personal boundaries. The last few years, many a book and article has discussed the subject of paradigms. A paradigm is a way of thinking or a way of viewing how the world is or how the world must be.

I once viewed a video that was about paradigms. An example they used in the video was about the Swiss watch makers in the 1970's. It seems the Swiss controlled 90% of all watch production in the world. And the other 10% was spread out amongst all the rest of the countries. But as you know in the late 1960's and early 1970's digital technology began to come alive. The Swiss were presented on a number of occasions with opportunities to get into the digital watch market.

But their paradigm wouldn't allow them to see it. You see, those watches didn't have mainsprings. They didn't even have hands or gears. That couldn't be a watch! Watches had always had those things. They couldn't imagine that people would go for this electronic gimmickry. Well, history certainly proved them wrong. Their paradigm trapped them. They could not see "beyond their boundaries." By the time 1990 had come around, the statistics had completely flopped. The Swiss control only 10% of the watch market in the world today. The other 90% is controlled by the rest of the world and the majority of that is Japan.

The above example really causes one to think about the entrapment that paradigmatic thinking causes. I'm sure you've seen examples of individual surveyors with paradigms that kept them from seeing beyond their personal boundaries. I remember many limitations (self imposed limitations) back in the early 1970's as electronic distance measuring devices were coming on the market. Similar things are taking place even today with GPS. Some of us just have a hard time seeing beyond our boundaries.

This chapter is not really about stepping into new technologies or even new markets (such as GIS). Rather, I would like to focus on our professional boundaries. Surveyors can be a very independent lot. We can also get terribly stubborn at times, can't we? Those kind of actions and attitudes indicate an unwillingness to step beyond boundaries. Both our personal and our professional boundaries are well-meshed together. We often cannot see beyond ourselves and our own little world in which we operate.

Where the survey profession needs to see beyond it's own boundaries the most, is in involvement in it's own profession. We have a terrible track record of not supporting our own profession. Based on some general figures that I know of, I would say less than 10% of the registered surveyors in the United States belong to ACSM![2] This figure is considerably lower than many other professional organizations. Numbers seem to increase somewhat for membership in state surveying associations, but are still woefully low if you want to call that association truly a representative of the profession.

Over the 10 years that I have conducted seminars, at almost every seminar, I have handed out a course evaluation sheet. This form gives people an opportunity to critique the seminar and make other general comments. At the head of the form are a few demographic questions. In particular, it asks if the person is registered and who paid for the seminar. But it also asks, "are you a member of ACSM?" In the past 10 years I've had 8,000 people attend my seminars. Of that crowd only 8% belong to ACSM. Similarly, of this same population, only 14% belong to their state surveying association.

---

[2] ACSM-America Congress on Surveying and Mapping; the only national surveying society.

Those who attend the seminars are really fine professionals. I say that because I respect anyone who commits themselves to taking seminars and participating in continuing education in that way. Every one of those people gave up at least a day to come listen to me talk about some narrow little niche of the surveying world.

So in essence I feel like I get to meet the "cream of the profession" at these sort of opportunities. I ask myself this question, "Why are these fine professionals willing to pay anywhere from 100 to 200 dollars each (depending on the length of the course) to listen to some guy they may have never heard of before?" And since they are willing to do just that, then why is it that so few of them feel that an equal amount of money would be worth joining their national and/or state organization? It is an important question that both the national and state associations need to answer. I'm not sticking my head in the sand and pretending that there aren't problems with ACSM (NSPS) as well as some of the state organizations. And I've belonged to a few state organizations where I felt like it's more of a country club. But I still belong because I want to have my say. Membership in those organizations provides too many other valuable opportunities for me to give it up just because of the politics.

But it's so easy as a surveyor to say, "I'm not spending 150 bucks belonging to that group." And we can justify that quite easily. But this is part of "seeing beyond our little boundaries." We must step beyond our little world. Our profession needs our support. We cannot let a small minority continue to dictate to us or represent our profession. And we cannot afford to keep ourselves in the dark within our limited paradigmatic boundaries, never to be involved in the profession as a whole.

I'm endorsing membership in professional organizations. I'm also endorsing participation. There are plenty of members in those organizations, but very few who participate. I beg you to participate in your professional association. Give it your support. Show up for special activities as they fit your schedule and lifestyle. Support things you personally may not even need.

I'd like to give an example of commitment to the profession. Over the years doing my seminars I have been shocked to see a few specific persons out there in the audience. They were people such as my friend Jerry Broadus of Washington State. I was honored to have him present. However I felt like he could teach the course just as well if not better. But many people like Jerry set a powerful and positive example for the rest of us. They go and participate in activities. They go and sit through perhaps some really boring seminars or some terribly depressing meetings. But they participate. They show up. They are there. That is the kind of spirit and attitude we need. That is what will make this profession rise in every aspect.

There are so many things that each of us can learn from one another, so many ideas. Going to association meetings has provided me with incredible opportunities to meet people, to pick up work, to get ideas for marketing, to solve problems that I'm facing on a particular survey. I can't say anything but good about those kind of situations. Why would anyone turn their back on those opportunities?

So can you see beyond your boundaries? Reach out and become a part of this profession. The ethical surveyor cannot afford to stay inside his little world. We must be a part. We can make things happen.

Chapter 14

COGO SLAVES

"People who like this sort of thing will find this the sort of thing they like."

Abraham Lincoln

With the arrival of the handheld calculator and the "PC" back in the office, the surveyor now can hold the tremendous power of coordinate geometry in the palm of his hand. COGO (coordinate geometry) is a very simple and easy to use process for computing positions of corners, lines, and boundaries. There is nothing wrong with that tool or the technology to use it. It's the mathematical backbone of a survey operation.

But as with any scientific technological advance in our profession, there is a tendency to abuse it and misuse it. While COGO should be a tool that is secondary to other factors in our survey process, it has in many ways become the focal point of our process. The ability to manipulate coordinates through adjustments and various analysis has become an enslavement to our profession. If you have a wild imagination, COGO can take you to any place that you desire. It is a tool that can completely hide the truth from you. It can make you feel good about yourself in spite of reality. After all, any error incurred can always be adjusted out. (See the chapter entitled compass rule).

The true COGO slave may or may not be registered as a surveyor. Their sole connection with surveying is through the manipulation of coordinates. I've met many of these people over the years. Some of them go by the title of, "plat reviewer" or "plat checker" in a county surveyors office or similar authority. Many others are just everyday surveyors trying to "make it fit".

Don't get me wrong. I'm all for good closures on surveys. I'm all for a plat that represents a true picture of the survey in the field. But what has happened to the COGO slave is a dangerous thing that we should all heed. One can become so wrapped up in "precision" that they totally miss the boat.

For example, I know of a couple locations where plats have been thoroughly reviewed. They were "redlined" so bad they looked like somebody had bled on them. The COGO slave insisted that the plat be adjusted by a hundredth here, or by a second there. This, that, and the other minor little change must be taken care of. Yet the plats themselves were totally bogus regarding the survey they represented. The procedure was completely improper but the COGO slave approved the plat once it closed perfectly. Is there something wrong with this picture?

It seems to me that our profession has lost sight of reality. The measurement requirements of a surveyor in the field to provide bearings to the second, distances to the hundredth, and areas to the ten thousandth, are beyond what most surveyors are regularly doing. When I see people using COGO to inform other surveyors of small "errors" in their survey, I get quite angry. The survey they compare to is not capable of producing the precision the numbers imply.

I had a surveyor call me one time to point out some errors in my survey. I had run a subdivision line and set pins every 100 feet or so on that line. These lot corners were set in 1980. In 1990 he informs me that my lot corners are off the line. Some are .02 east, some .04 west, some on line, and others within .01 feet. I asked him how he determined these "errors". He had set the instrument on a hill some 1000 feet from the line, and radially shot them in. The error in his survey, from both instrumentation limitations and operator bias would exceed what he was telling me I was off. Yet he insisted I was in error. He used COGO to "inverse" between my positions and his pure calculated positions, and sure enough, there was .01 feet in there!

The above scenario is repeated thousands of times daily. It should do more than make you sick; it should scare you. A person with a measuring device and a COGO program can make almost anything seem precise, but it is all out of focus if we do not understand errors, uncertainties, and instrument limitations. I've met many a survey crew in the field that has no idea what a prism offset is, but will insist that your corner is off by a hundredth or two.

One may ask, how is this an ethical issue? The Professional Surveyor exists to provide certain professional services to the public. These services require special training, experience, and skill. That is why we are licensed and regulated. That is also why the public expects us to be the experts in these areas. We hold ourselves out as the measurement experts of the world, yet we lack understanding of our limitations and how they fit into reality. We've allowed the number of places behind the decimal on our LED readout to deceive us into what we can really produce and provide. We are lying to the public with overstated accuracies. We have played COGO games with ourselves and our fellow professionals and think we are doing something good and right. In reality, we are shaming an honorable profession.

The heart and core of ethical issues is honesty. COGO is a wonderful tool. It can also lead us down a foolish path of surrealistic expectations of ourselves and others. It is time to wake up and know the limitations we have. I strongly urge surveyors to take some training in error sources and analysis, estimated uncertainty, and basic statistical analysis as it applies to surveying. You will be shocked at what you should be claiming on your plat. You will realize that the misuse and outright exploitation of COGO is an embarrassment to our profession. The ethical surveyor must realize the folly of using precise numbers without benefit of precise measurements.

Chapter 15

CONTINUING EDUCATION

"Learning isn't a means to an end; it is an end in itself."
Robert A. Heinlein

By mid 1995 a dozen states have enacted mandatory continuing education laws or regulations. This author has not been silent on his support for mandatory continuing education.

The cynics may suggest that I would support continuing education primarily because I've been in the continuing education business. But I have to tell you; the reason I got into the continuing education business was not to make money. Rather it was to provide a quality service I felt was greatly lacking in the profession.

Continuing education by it's name suggests that there's a need to continue learning. That doesn't mean there's always new things coming up, although with technological changes, computer software changes, and even legal changes there are many things for a surveyor to keep abreast upon. But it's this authors opinion that the average land surveyor cannot be an expert in every possible field within land surveying. We'll discuss that more in another chapter under the subject of Specialization.

But do we really know everything we think we know? That's the question we have to ask ourselves. I have personally found that continuing education is a major boost to my professionalism. My attending seminars, workshops, and conventions has really opened my eyes to the many different things I didn't know. I have learned far more since being licensed than I did before.

The surveyor who vehemently opposes the establishment of mandatory continuing education laws really scares me. I'm not sure what their motivation is. Perhaps their concern that "Big Government" is intervening. I have to tell you, I hate "Big Government" and I hate Government intervention in most anything. But when a profession will not police itself and will not take care of it's own problems; that's when you invite Government intervention. And that's where we're at.

We could police ourselves. Your state does not have to pass a law if your profession were policed through the society and it became an important, integral part of being registered in that state. If we all had a participative attitude it could pay off. But the bottom line is many surveyors, if not most, are fiercely independent and want to exercise their "rights", even if that right is the right to be stupid.

Some surveyors feel they really don't need continuing education. They already know what there is to know. Those are the ones that scare me the most. They're the ones that I have met over the years at various functions who spend their time shooting the bull, drinking, or heckling seminar speakers. More importantly, I've run into their work on various projects and I've absolutely been staggered at their lack of basic knowledge of fundamental principles of land law.

Our profession is literally filled with people who do not know the basics. Some were "Grandfathered" in or entered the profession back when the survey test was a simplistic approach. Some are those of us that passed the 16 hour NCEE examination like myself. We do not know all there is. There is far more to be learned. The arrogance of someone who would suggest that they already know everything they need to know staggers my mind. Actually, any such person should be investigated immediately; I'm sure you'd find enough evidence to get rid of their license.

I also realize that there are some who just complain about the cost of continuing education. In reality the cost is extremely minimal, especially compared to the knowledge you might gain and the liability you might be able to reduce simply based on your increased knowledge level. The cost is minimal. We forget to look at the overall benefits.

I realize that one of the great complaints about continuing education is that there's not that much good stuff available. That's why I got into the continuing education business to begin with. I tried to offer a very different approach and a very specialized niche, that being the Public Lands Survey System. And I realize that many of us have been to many a convention where we heard the same speakers saying the same stories. I've been to several seminars, paid good money for them and heard the exact same speaker say the exact same thing he had said 10 years ago.

I believe if you create the market for continuing education you will create far more opportunities and the market will respond to that. Now, there will be some real Bozos out there, I don't question that. I've been to some of their sessions too. People who neither have the knowledge and background to be teaching nor any idea what it means to teach.

I agree that there should be some form of standardized teaching credentials. I would suggest that this be more on a National level. Speaking from the educators point of view I really don't want to have to go through 50 different processes to justify my credentials as a speaker. This would be an area for ACSM to lead the way and for the states and their state societies to all acknowledge and accept their leadership in that way.

If you really want to help your profession, if you really want to help clean it up, if you'd really like to get rid of the bad practitioner, continuing education is one of the strongest ways you can do it! It allows you a tool to force everybody into a wide variety of educational opportunities. They will be exposed to the facts. Once they've been exposed to the facts but continue to screw up out there in the field, I have a lot less conscience problem going after them and nailing them for their license.

You will also find, as some of the states have done, that when you enact continuing education some people who aren't wholeheartedly in the surveying profession will drop their license rather than get into that additional expense and trouble. Good! I have no complaints about that. That should be one of our goals. But the chief goal of continuing education is to upgrade our entire profession. It embarrasses me when I see that in almost every state, realtors, real estate salesmen, are required to do continuing education. I have taught many of their continuing education classes. And that's opened up a lot of opportunities for me to open their eyes to some basic surveying principles of which they have no concept, but are useful to them. It really embarrasses me when I see realtors being required to take it and then surveyors just battling it out like it's the end of the world. In reality it's the beginning of a major step towards professionalism. The ethics of avoiding such a powerful tool cannot be dismissed.

Every state needs mandatory continuing education. Every surveyor needs to be exposed to a wide variety of subjects and viewpoints. The truly ethical surveyor will participate anyway. Let's upgrade our entire profession by bringing every registrant into the fold. Kicking and screaming may be their initial reaction. Keep an eye on those who react that way. They are the ones bringing down your profession in many other ways than continuing education.

Chapter 16

I SAW YOU THERE

"Life is like riding a bicycle; you don't fall off unless you
stop pedaling."

Claude Pepper

At first I did not notice you. You were sitting in the very back row of seats and it was difficult to clearly see you. But you were there.

I was teaching a class on a very important legal subject at your state surveying convention. It's not that I was the best speaker on Earth or anything, or that my presentation style was perfect. I was doing my best for about three hundred others who were there with you. You had paid good money to hear what I had to say. The leaders of your state's professional association felt my subject was vital to the upgrading of your profession. And I saw you there.

Your state had recently passed a mandatory continuing education law. Every surveyor in your state had to attend a certain number of hours of courses to renew their license. It was a bold and professional move on their part. And there you were.

The vast majority of your colleagues were paying close attention to my words. They were copying down the drawings I had made on the board for examples of the common but complex issues we all face. Your friends and fellow professionals were soaking up what they could use for the improvement of your profession. They eagerly wanted to learn, and that excitement spread to me as the speaker. There is a synergistic effort when everyone in the room gets involved. Some wonderful things were happening for all of us.

Except for you. I saw you there. In the back row. You were reading the current issue of "USA TODAY". You kept reading it through most of the morning. You took no notes. You did not open the lengthy handout I had provided you. You never looked up to see an example. You never asked a question. You just put in your time.

It did not distract me. It did not seem to distract your colleagues either. But you

48

embarrassed yourself. You were there only to get your "credits". You did not give a damn about the subject. I doubt you gave much thought to anyone but yourself that day. You certainly showed your worth to our profession.

During the afternoon a State Board member informed you that your behavior was not going to earn you any credits, and you stormed away from the meeting. I'm sorry if it was my style, subject, or approach. I'll try to do better next time. But the least you could have done was show some token interest in your own chosen profession, and in your fellow professionals. How do you think your haughty attitude came across to those anxious young professionals sitting next to you? I felt sorry for you; and for all your clients. It was a poor showing on your part. Your arrogance was only exceeded by your ignorance.

You are the type of person mandatory continuing education is made for; the licensee who thinks he knows it all, or figures he does not need to know anything. I think I saw a distant cousin of yours at another session I did in another state. He was so bold as to ask for the completion certificate after only one hour of an eight hour course. He wanted to prove he had been there, but was planning to leave early. I knew he had to have been related to you somehow.

I was just the speaker....and I saw you there. But most importantly....we all saw you there. And surveyors like you really scare all of us.

Chapter 17

DEAR ACSM

"Where there is no vision, the people perish."

Proverbs 29:18

My membership in the American Congress on Surveying and Mapping (ACSM) has come and gone several times. Sometimes it was because I just couldn't afford the membership dues. Other times it was because I was angry with them and some of the unethical things that had gone on in NSPS (which is a member organization of ACSM). Still other times it was just a matter of complacency. Yet I still strongly support participation in our profession.

As mentioned elsewhere in this book, I have canvassed a large number of surveyors (over 8000), all of whom have taken seminars from me. These people are generally not members of their state professional organization. Only 14 percent of these people I have canvassed belong to their state association. But it gets worse, less than 8 percent belong to the national surveying organization, ACSM.

ACSM/NSPS has the unenviable task of being the representatives of the surveying profession. The reason that is so unenviable is because leading a group of surveyors is an almost impossible task. In fact, just trying to organize surveyors into a meeting can be an almost impossible task.

Surveyors are very independent. It is one character trait that attracts most people to surveying. While admirable, it prevents surveyors from fully grasping the concept that there is strength in numbers. Only marginal numbers have maintained their membership in ACSM, however.

While I do not know the specific numbers, I suspect that less than 10% of all registered land surveyors in the United States belong to ACSM. This is the only national organization that specifically represents surveyors. Trying to lead and organize surveyors is like trying to get your cat to do what you want it to do. Cats are quite aloof; they don't really pay any attention to you, act dumb sometimes and other times do the opposite of what you just told them not to do. Surveyors are like that.

So, this is what I'm saying: Surveyors are a tough audience. There's rarely a time when surveyors agree on any subject and usually they have a diverse set of opinions about even the simplest of concerns. So I do not blame ACSM for not being able to get a 100 percent participation. It is impossible, given the nature of the character of surveyors.

But I do have some serious concerns regarding ACSM and the member organization NSPS. There has got to be some significant reasons why surveyors are so poorly involved in their national organization. I would like to offer a few thoughts as to why this might be, with the hopes that ACSM and NSPS would hear and at least consider some changes.

Before I proceed, I want to note that most of the individuals whom I have met who were officers of the national organizations have been some of the most marvelous professionals I have ever met. These comments are not meant in any way to be derogatory to any of those individuals. In fact, I salute many of our past presidents and area directors who are either acquaintances or very good friends. I believe that you have tried very hard to make some significant in-roads in many ways on our behalf. And I know what a thankless job it is when a vast majority of the profession don't get involved or even belong. So I pause here to thank all of you for your efforts, time, and concern for my profession.

So, what might be the reasons surveyors are not involved? I think there are four key reasons:

- Surveyors lack vision — they fail to see the big picture.
- Surveyors are cheapskates.
- Surveyors are frightened away by how they perceive the national organization.
- Many surveyors don't know ACSM exists.

Let's discuss these four in a little more detail.

The little mom and pop surveying organization has it's own perception of the world. In particular it is desperately seeking out work, dealing with surveys and projects that constantly need extra attention. Just getting the product out the door can be very trying. In the United States, the majority of registered surveyors work for an organization of less than ten employees. That's significant. We must realize that surveyors are generally located in small organizations working in relatively small geographic areas.

As most of you know, it is easy to get your nose buried in your local community. The day-to day grind of keeping a small surveying organization going can cause one to lose sight of the big picture. If the majority of your experience as a professional surveyor has simply been going out and cutting the brush, looking for the stone, and drawing the plat, it's hard to realize that there is a group of over 500 people in a big building on a hill in Washington D.C. who are making decisions that affect your livelihood.

Maybe the decisions Congress makes never directly affects your livelihood. Or more likely, you've never realized how those decisions they've made have already affected your livelihood. One of the things that ACSM has done in the past is to represent the surveying profession in various legislative actions on Capitol Hill. As we move into the high-tech world of geographic information systems (GIS) and other landnet-based systems, we will see more legislation going on in which the surveyors interests must be protected. I think there have been some bold actions on the part of ACSM in this regard in the past and I thank them for it.

There really is a big world out there. International issues are now facing the local surveyor. Actions such as NAFTA and GATT will have tremendous impact on the surveying and engineering business in certain parts of the countries that are affected. We must recognize that our society has become a global economy. It will impact our lives.

So surveyors, we cannot afford to lack vision. We must see the big picture and the big picture means we must unite and be involved.

The second reason mentioned above is that surveyors are cheapskates. I don't mean that to degrade any of us. However, the vast majority of surveyors that I personally know are pinching pennies. It's not necessarily just because it's their nature, but rather because it's a requirement. Surveying fees have not kept up with inflation, let alone with the complexity of the work involved. And surveying can be a roller coaster business.

We are very much tied to the economy, especially the construction side of the economy. So we ride the waves of "boom and bust" just like many a contractor or engineering firm does. One year, you're so busy you can't think straight, and the next year you're crying for enough work just to pay the wages.

Surveyors are also not well known for being good business persons. And it would seem that we often do not save and manage our money in a wise way. Managing a surveying business, paying attention to the technical issues, saving for retirement, and handling all sorts of personnel issues is not what we do best, nor what we want to do!

Whatever the reason, these financial pressures on surveyors have caused them to be quite cautious with their money. Nobody's going to throw away $150 to $200 on something in which they see no personal value.

The current annual dues in ACSM and NSPS are about $150. In reality, this is a very small sum. Yet surveyors are not willing to part with that. I believe that ACSM really needs to reduce those rates. Reducing the income is going to require reducing the outgo, which I will discuss in the next section. Suffice to say that psychologically speaking, ACSM has priced itself beyond the reach of the average surveyor. I say psychologically because it is the perception that $150 goes into this national organization for which very little is perceived to come back. I would prefer to see ACSM reduce the dues and perhaps offset that at least in part with an increase in membership.

Personally, I feel it would be better if every registered surveyor in the country belonged to ACSM and paid $50 a year, rather than less than 10% of us paying in $150. I also realize that we will never have 100% participation.

Some of the things that ACSM offer's its membership are really of minimal use or interest to the average surveyor. While I have appreciated a few articles written in the

ACSM bulletin and other publications, I have found that less than 5% of those articles have any direct relevance to me in my surveying practice. This is not to say that the articles are not of value to someone, I'm sure they are. There's usually very high quality literature in the magazines. But no matter how well written an article is, and no matter how much it is right to the point, if it only interests 2 or 3% of the membership, it has missed the mark.

Further, the membership flier is clogged with a list of benefits, but are they really the things we are looking for from a national professional organization? Rental car discounts, that are not as good as the discounts most of us get for belonging to a dozen other groups, do little for me. Neither do discounts for things I'll never own, never use.

The third area I want to mention is changes in ACSM. ACSM is perceived to be a rather fat and complex organization. I would tend to agree with that analysis. They have reduced their overhead in recent years, and this is appropriate. But I still do not fully understand how ACSM and NSPS are structured. It seems to me that they have an organizational chart more complex than the United States government. I was never quite sure who represented me and who else that person represented. I never quite understood who could vote on what, and who goes to what and who does what, or who said what.

I am personally still very frustrated with ACSM. I have been amazed at the number of people that they fly (with our dues) back and forth across the country to these meetings and national conventions. There are committee meetings and sub-committee meetings. I wonder, "what value am I, a general member, getting out of those meetings?" I don't even know what half of those groups do or talk about. There are no reports available.

Another frustration I've had with ACSM is the convention schedule. This is one of the few professions that I know of that has two annual conventions. Almost everyone else has one convention per year. Yet we continue to see two conventions annually and we scratch our heads wondering why they often lose money. We need to wake up here! People can't afford to go to two. In fact, if you're lucky, you'll get someone to go to one, every ten years. So why in the world would we continue having two conventions a year? It is overkill in my view. Let's let a little time pass so the exhibitors can bring something new, so the presenters can take a breather, and the members save a little money for the next trip. Halving the number of conventions may even double the number attending.

ACSM is perceived to be controlled by members on the east coast. This is essentially true. There are a couple of reasons for that.

- There are more surveyors on the east coast than there are in the rest of the country.
- Eastern surveyors seem to be more involved in ACSM than those in the west.

The greatest drawback to this situation is that those who feel they are not represented continue to withdraw from the organization. ACSM is aware of this concern, but perhaps they could do additional things to alleviate the worsening condition.

A final area to mention is that of awareness of ACSM's existence. I remember walking along the beach at Virginia Beach, in Virginia, with a fellow U.S. Forest Service surveyor. We were in town to attend the ACSM convention. We ran into some surveyors working along the shoreline. We asked them if they were attending the national convention. They replied that they did not know it was there, nor did they know what ACSM was.

I was in the surveying business for 11 years before anyone ever mentioned a national (or state for that matter) surveying organization. It seems to me the word has not spread very widely.

Perhaps we can find better ways to "market" our associations. We need to push the current ACSM members to spread the word. Perhaps a special (greatly reduced price) introductory membership would be in order. Could we get names of new registrants from each state and send them a package? Could we use mailing lists from the two national surveying magazines to contact more people? Could we give dues discounts to members who bring new members to ACSM?. If a person brought in, for example, three new members, wouldn't this be worth free annual dues?

In conclusion, I offer these basic suggestions to ACSM.

- Simplify your structure.
- Reduce the convention schedule to one per year and continue to move those around the country as has been done in the past.
- Reduce the dues and go on a massive marketing campaign to get people involved at a greatly reduced fee.
- Strive to improve the publications so that they represent issues of concern to the majority of surveyors.
- Look for new ways to get the word out on the existence and purpose of ACSM.

I do not want to seem arrogant or naive in this issue. I hurt as a professional surveyor that our national (and state) organizations are so poorly accepted. These thoughts are offered only because of my concern about our profession, and our own self-perception of the profession and the world around us.

To quote the Book of Proverbs, "Where there is no vision, the people perish". Surely we are a profession that lacks vision....less than 10% membership in ACSM speaks for itself. If you are not involved now, please change it. We need you.

Chapter 18

CONFLICTS OF INTEREST

"A situation in which regard for one duty tends to lead to disregard of another."

Black's Law Dictionary

One of the more important ethical issues any professional should consider is that of the conflict of interest. A professional should maintain an "arms length" relationship from certain projects or activities. A conflict of interest is where you have more than a professional technical reason for being involved in a project or venture. We cannot afford to degrade the profession in such a way as to be exposed having interest and motivations that go far beyond our professional input and duty.

A more difficult concept is that of the "appearance" of a conflict of interest. This term takes one beyond an actual conflict. It refers to the possible perception by others that there might be a conflict, even if there is none in reality. Most conflict regulations in government, and guidelines in private industry, refer most often to the "appearance". The difficulty with this concept is complex: Anyone can find an appearance of conflict in almost any activity. In this legalistic world we live in, you can hardly show up for work without there being someone who thinks you've done so for ulterior motives.

One major form of conflict is the subject of moonlighting. That topic is discussed in it's own chapter in this book. But as a review, the moonlighting employee can really find a conflict when working in the same line of work for another firm or for himself. Both the primary employer and the moonlighter have serious potential conflicts, if not some in reality.

Another form of conflict we must consider is that of surveying your own property. On the surface, one would be insulted to think that others would accuse us of changing our survey results for our own benefit. Believe it or not, that has happened. But the "appearance" issue raises it's ugly head on this sort of problem. By surveying our own property we simply invite more trouble than it is worth. Everyone will be suspicious. You will spend more time defending a good survey than you saved doing it yourself anyway. Avoid this conflict at all costs. The attorneys say "any attorney who represents himself has a fool for a client." We could learn from this. (See chapter 47, "Do It Yourself".)

What about the surveyor who has a financial interest in a project? This carries similar baggage as the previous discussion. Say you are surveying a subdivision in which you have a financial interest. It is not illegal in most jurisdictions, but it definitely creates the appearance of a conflict. Similarly, the surveyor who gets government contracts on a regular basis, where the Contracting Officer is a relative, may have much to answer for, even if all is above board.

Finally, be aware of the dangers of working on a project which adjoins one of your previous projects. In 1990 I knew of a surveyor who took on a section subdivision project. It was in a section next to one he had worked in two years before. During the course of the work he discovered a significant error in his placement of the quarter corner common to the two sections. A vast amount of construction had taken place based on his first corner position. If he were to accept the newly found evidence of the original corner, he would place three buildings and a street in trespass. He chose to destroy the original evidence and leave his erroneous position to conceal his earlier error.

Whenever we work against our own work, we must have a willingness to face our errors that is identical to our dealing with the errors of others. This is a tough situation but I can see no choice. The surveyor described above thought he'd gotten away with it. However, I had just tied down the same original evidence two weeks before he destroyed it. I knew it existed, and was about to file a plat showing it. He was caught not only screwing up the first survey, but destroying evidence, covering his butt, and lying to the public through a bogus plat. The gentleman is no longer licensed.

If you let your imagination run, you can invent millions of "appearances" and conflicts. You cannot hide from every shadow out there. The prudent professional will have his "ducks in order" on these issues. The way you do that is by thinking ahead of time about the possibilities, and taking appropriate action to diffuse them ahead of time. Above all, be sure your ethical responsibilities to the profession and society have been considered. Seek advice. Get input on certain issues. The important thing is to think. In the end, it will be your view of ethics against that of others. Be prepared!

When in doubt, don't do it. But remember that a lack of action in some cases could be considered a conflict of interest in itself. If you see an error in your work, or that of others, you must deal with it. Hiding your head in the sand never works out for all concerned.

Conflicts of interest are waiting for you. Keep your ethical thinking cap on and you'll do well.

Chapter 19

MOONLIGHTING

"Hey, what's wrong with earning a few extra bucks?"
Anonymous

In the last five years I have conducted seminars on the subject of ethics for land surveyors. Of all the possible topics, there has never been a subject that has produced as much controversy as the subject of moonlighting. The term moonlighting means different things to different people. There are some hot emotions both for and against this subject.

Let's define moonlighting. Your dictionary will define it very simply. It will say something like, "A job that one holds in addition to his regular job". In the broadest sense that truly is what the word means. Some surveyors have more specific and well defined definitions of moonlighting.

There are many in our profession who feel that it is totally unethical for someone to moonlight in any way. For instance, they feel that it is improper for some of their help to work at Seven Eleven on the weekends.

It's important to remember that in some parts of the country we are not able to pay even our best help adequately. Many of them are forced by basic economics and other personal circumstances to be in a moonlighting situation. Obviously, the concern of the employer is that an employee needs to be in top shape and able to provide full performance on their regular job.

Of more concern to surveyors is the person who is moonlighting doing survey-related work. This would include people who are surveying on the weekends for another firm. Or perhaps they are working with a registrant in the parent firm on the weekends doing side jobs. I've known of people who have used their drafting skills and their CAD skills to work on the weekends or evenings to try to pick up extra money. It's hard to fault these people for doing that sort of thing.

Still there are others whose moonlighting has been a great benefit to the profession as a whole. I would remind the reader that virtually every surveying text book and virtually every surveying seminar and workshop has been done by someone moonlighting. I do not

57

know of anyone who's full time job is simply writing surveying books or giving seminars or lectures.

There are even surveyors who are contributing to our profession and our communities by teaching surveying on the weekends or evenings to provide educational opportunities at the community college level for future surveyors. Some of the most significant seminars or lectures that I have ever attended were done by people who did not teach full time. In fact, their real-world work is what makes their teaching far more valuable than that of some egg-head professor who never had a real surveying job. Our profession can benefit from certain types of moonlighting.

It seems to this author you cannot draw an absolute line regarding moonlighting, and yet there are some very serious concerns.

Moonlighting, of course as mentioned above, has provided additional income to many a person working in the survey related fields. There is a positive benefit to that but one must weigh this against the negatives of the employee whose energies may be diverted.

Perhaps one of the most significant things over the years from actual field work type moonlighting is that many a surveyor has got his start that way. A number of surveying or engineering businesses that you know have been around a long time, actually started with the original registrant moonlighting. This often occurs just after receipt of the first L.S. license. The surveyor finds him or herself in the position of needing to pick up work to provide additional income. He needs to buy instruments and other equipment. He wants to start to build up a clientele. I personally know of several people who are adamantly against moonlighting, but I know they got their start through the very process they now condemn.

There are some very serious, negative sides to the subject of moonlighting also. The first of these is the question of loyalties. When the surveyor is working for one employer as his forty hour-a-week job but then is working for another surveyor (or even himself) in the evenings or weekends, a serious question of loyalty arises. Many a surveyor has found himself fired from his principal duties due to the fact that he was caught in a conflict of interest. Surveyors have diverted work from their chief employer to themselves. Even worse, surveyors have stolen major clients from their original firm. These are certainly unethical and improper forms of behavior.

Further, the interests of the employee could negatively impact the main employer simply due to effectiveness. The employee who has moonlighted several nights or all weekend may be of as little use to the employer as the hung over employee. An employee has certain obligations to the main employer, and these must be met!

Another concern is the over all effect on the economy of the profession. Many a "back-door" surveying organization operates and undercuts the well established firms.

People who do surveys with little or no overhead certainly do a disservice to firms that are in the business on a full time basis to provide the same surveying services. Some have argued that there's no requirement that you have huge overhead, that you be a large overly burdened organization to provide surveying services to people. I certainly agree with that. However, the entire idea of getting a professional survey the same way you might get some plumbing because you know a guy and he'll come do it on the weekend certainly has a negative impact on the image of the profession as well as the income.

One of the universal hatreds the surveying profession holds or shares is that of the Government employee who's moonlighting on the weekends. There's a special concern about the surveyor who operates in a public or quasi-public capacity and yet is operating privately on the weekends. The concerns usually revolve around the issues of whether the person is getting work through the contacts that he has in his regular Government job. Further, there can be the appearance of a conflict of interest for any government employee of any government level to be surveying on the weekends.

Of great concern with the moonlighting government employee is the "appearance" that the government entity he is regularly employed by has somehow sanctioned or approved of his surveying. There is even an appearance that the manner in which he surveys has been sanctioned by that government entity. This concern, in fact, can be expanded to the non-government employee. There are situations where people moonlighting in other professions have made errors in their work and the parent organization that they are employed by was held liable and responsible. This is a function of what the law calls the "master servant relationship". You would be surprised how far reaching that relationship can extend, especially in the area of liability and moonlighting.

The employer who not only knows of the moonlighting actions of employees but also allows them to continue is essentially condoning the action. This issue should be of concern to all of us; not just the government employee.

As you can see there is tremendous potential liability for the private surveying organization to allow it's employees to moonlight. At a minimum a firm should contact their attorneys and develop a strict moonlighting policy in the form of a handbook that would regulate such actions. One needs to control or perhaps forbid any moonlighting altogether. This would be a wise and proper action. Most government agencies require written permission before an employee in any profession is allowed to moonlight. This would be a good practice for the private firm. It is an ethical obligation on the part of the employee to inform his employer of any conflicts of interest.

This author has been criticized and accused on a number of occasions because of his Federal employment during a portion of the time that he has done surveying and seminars. I will point out that I have always had written permission from my employer. However, that does not eliminate the appearance of a conflict of interest at times and it has certainly raised the concerns of a number of people both in and out of the government.

My writing of this book is moonlighting! I do not write these things to defend myself but rather to point out that these ethical issues come close to home constantly.

So what's the bottom line with moonlighting? Moonlighting creates many difficult situations for our profession as well as the individual. We owe a loyal and thorough service to our main employer. When these duties are impaired by outside activities, we have crossed an ethical line. When these duties create conflicts of interest, the primary employer's image should be of utmost concern. The appearance of a conflict of interest should be considered.

Truly, the employer and employee should discuss thoroughly and openly any moonlighting activities, existing or proposed. The employee has a heavy responsibility in this regard. The employer must consider liabilities, image problems, and difficulties with implied approval when viewing moonlighting.

Moonlighting is acceptable under certain circumstances. The arrangement should be above board and in writing. This will protect all parties concerned. Any outside activity which can or will create significant difficulties for the primary employer should be eliminated.

Whatever you do, do not take this subject lightly. This sort of subject is at the heart of ethical concerns.

Chapter 20

THE COMPASS RULE

"Captain, this is not logical."

Spock, on numerous occasions, Star Trek

As we have already seen with many of these subjects surrounding ethics, the concept of "honesty" is ever present. It's important for surveyors to realize that they must apply the principles of honesty to many things besides just telling the truth in a sentence they may state. There are in fact, many survey plats bearing bald-face lies. I'm of course speaking of measurements.

The last 25 years in our profession have seen major changes in our ability to measure. In fact, the technological changes we've seen have greatly improved the accuracy and efficiency of our survey crews. It has reached the point where many surveyors are operating with two person crews rather than the old three person crew simply because the technology allows us to do it.

With this technology and ability to measure we also have the appearance of accuracies. I say the "appearance" of accuracies for a reason. I don't question that over a long distance, over rough terrain, an E.D.M. can measure more accurately than what we might have done with a chain. But the devices we are using sometimes imply greater accuracy than they are actually able to produce. Further, our ignorance of the processes and limitations of the equipment can create a false sense of precision.

For instance: the E.D.M. that reads to the thousandth of a foot. All you have to do is go to the manufacturer's specifications and you will see that the ability to measure that distance is limited. The specifications themselves prove that the "thousandths" of a foot figure is erroneous and meaningless. Further, with a CAD program (which will be discussed in a separate chapter) we now have the ability to use coordinate geometry programs and produce computed bearings and distances with readouts that imply tremendous precision.

But is all of this realistic? Do we really think we can measure bearings and distances for the standard land survey to the nearest second and the nearest hundredth of a foot? It is this "slavery" to measurements and the "selling of our souls" away from the legal principles of land surveying that has created so much of the double monumentation we find.

Our measurement wars show up on the ground every day!

Most of my career I have worked in a rural environment. My typical traverse around a section or a large parcel of land would include a mixture of both long shots and shorter distances and angles in between. It's always been my desire to have a closed traverse. I realize many surveyors today do not do that. Most states have some kind of survey standards for rural areas. Generally that's a one part in 5000 closure. Hopefully most of us are getting far better closures than 1:5000 but think about it; if you traversed around the exterior of a section in order to control it for a section subdivision, you've traversed 4 miles. That means you're going to have four feet of error in your traverse. It's still acceptable according to the 1:5000 standard. Even if you close 1:20000, you still have over a foot of error in your traverse.

What concerns me is this: we are in the habit, for the last 200 years, of using what is called the "compass rule". The name of that ought to give us a clue as to it's best application. The compass rule assumes that the errors are proportionally distributed along the line. Therefore it will distribute the corrections based on the length of the line as it is to the entire length of the traverse. In essence you adjust the bearings and the distances of the longest lines of the traverse far more than the shortest. But when you consider the instrumentation that we have now-a-days that is exactly the opposite affect that you would want to have. It is my long shots and long distances (with good angle backsights and foresights) that are the strongest things in my traverse. The weakest points in my traverse are those little 100 and 200 foot shots that I have had to take through the woods through the mountains around all sorts of obstacles, unable to get long clean shots.

The compass rule does not apply to the technology we use today. In my opinion the use of the compass rule on a traverse is dishonest. It causes us to report different distances on the plat than what we actually measure. And usually it creates more confusion for surveyors who attempt to walk in our footsteps.

There are far better ways to adjust a traverse and to analyze it's closure. I personally would like to close about 1:30,000 and then just use the raw data. Unfortunately some of the states that I'm licensed in don't allow me to do that. They require me to adjust it by the compass rule. They require me to report bearings to the nearest second and distances to the nearest hundredth of a foot. They require it even though the instrumentation that most surveyors are using cannot mathematically support that kind of accuracy.

It is not the purpose of this chapter to introduce and teach other mathematical methods. It is however my purpose to encourage surveyors to honestly step back and see what they are doing. We must get out of the old paradigms of, "Run a traverse, adjust the traverse, and report unusable facts." We need to be honest with our measurements. I think there is a real ethical problem with applying a 200 year old (or even more) mathematical method to a technology and instrumentation that arrived in just the last 20 years.

Chapter 21

EEO

"We hold these truths to be self-evident, that all men are created equal, that they are endowed by their Creator with certain unalienable rights, that among these are life, liberty, and the pursuit of happiness."

Declaration of Independence

In the United States the Equal Employment Opportunity laws and regulations have been around for some time. Many other countries have similar legislation and standards to guide and control the employment practices of businesses of various sizes. The general intent of these laws is to ensure that people are not discriminated against in their ability to get a job based on their race, ethnic background, religious or political views.

It's a shame that the government has to pass laws to try to force people to treat others as they ought to be anyway. I recognize the reality of this issue. In spite of the best intentions of the liberal elements of our society, you cannot legislate morality. You cannot force people to get along when it's built into their nature not to get along. And therefore there are many places where the EEO Laws are ignored or brutally stepped upon. There are many injustices occurring everyday in violation of the spirit of these laws.

Land surveyors have traditionally been men. Much has changed in the last 20 years. We now see far more women getting into surveying as well as people of different ethnic backgrounds. In doing my seminars around the country I'm impressed with the significant changes that are taking place in the demographics of our profession. In particular, on the west coast I notice many Asian-Americans who are involved in our profession and are taking leadership roles in it. In fact, I've even met a few people who would be considered physically handicapped or disabled who are in the land surveying profession and doing an excellent job.

The spirit and intent of the EEO Laws and other related regulations is to give people a fair chance. Within the land surveying profession I've seen some strong resistance to the diversification of our profession.

The surveying profession has long been similar to that of other outdoor related professions. They seemed reserved for the "Marlboro" man type image, the rugged outdoors man, the rough and tough type. But times are changing; the whole world is changing and that stereotype is truly not applicable. As long as we're trying to live out that stereotype we will have a hard time accepting women into the profession. We'll also have a hard time accepting people who don't speak pure "American" into our profession. And so we see some discrimination against persons of ethnic backgrounds other than the white male.

I've had opportunities to meet many of the women in our profession and the vast majority of them are high-caliber professionals. Most are more devoted and committed to professionalism than the rest of us. It's truly inspiring to work with these people. They all have so much to contribute to our profession. Surely we should not ignore them.

It becomes an ethical question for the land surveyor. We may have personal prejudices that may have been built in to our childhood. We may have developed some intolerance for various ethnic groups. [1]

The truly professional surveyor, the truly ethical surveyor, must set these differences aside. We must be far more tolerant of difference.

This author has experienced a small amount of discrimination. That might surprise some reading this, especially if you know that I'm just your standard white male. But I have experienced something that many might not have and that is discrimination based on religion. In 1974 I was working for a civil engineering firm in Scottsdale, Arizona. I had worked for them for almost a year and had moved up in the organization and was well liked. I worked very hard and practically sold my soul to that company.

But after about a year I got a new boss. This new boss was apparently very intolerant of anyone that was different from him. It didn't matter how hard I worked or how nice I tried to be or how much extra effort I put in to getting along with everyone and producing what they wanted. It didn't matter. You see I go to church on Saturday and I have for 23 years. That's a little different than most people. For a 24 hour period I don't have anything to do with work. Rather, that time, by my choice, is set aside for the purpose of enjoying my family, relaxing in my home, doing things together with my family and participating in church services.   Because of this difference and a couple of related differences, this supervisor proceeded to fire me. He never had the guts to tell me straight up what it was about but all the other employees who knew what was going on did tell me. They understood it.

Back then I was rather naive and young. I didn't realize that I could have brought a

---

[1] It should be noted that it is most often not the color of skin or even language that offends us. It is most often the different culture of the individual with which we feel most uncomfortable.

law suit against them. I really didn't want to make any money out of the deal, although I was unemployed for some time thanks to this action. But what I really wanted to do was teach that guy and the corporation a big lesson. It's not worth running over one of your employees, especially a good employee, when the only thing you don't like is some little difference in their religion.

This same issue can be expanded to all of the differences we're discussing here. Surely we as a profession can rise above this unprofessional and unethical practice of judging people, employees or potential employees, based on things that have little or nothing to do with their job performance. Does the color of a man's skin change the quality of his survey? Only if you believe in stereotypes. Does an accent they speak with change the caliber of the measurements they can take? Does the deeply held religious beliefs of an individual get in the way of an effective, professional organization? Rarely!

So how about you? Whether there are EEO laws or not; whether somebody is looking over your shoulder or not; what's the bottom line with you? Are you tolerant enough of differences with people that you're still willing and able to have them participate in your organization? I tell you the truth, that organization that fired me lost one of the best employees they ever had! They were stupid as well as unethical! What about you? How do you feel about these things and what have you done in the past?

It's not a question of what does the law require. It's not a question of what can you get away with. It's a real question of ethics. What kind of human being are you? How do relate to other people, especially your employees or your potential employees? Rise above mediocrity and ignorance.

Chapter 22

WHO IS THE
NEIGHBOR?

"A professional is one who can do his best work when he
doesn't feel like it."

Alistair Cooke

All of us have experienced situations in the market place where prices are "jacked up" based on certain news items or events in a local area. Witness the fluctuations in gasoline prices at certain times. Witness the rates of motel rooms and airfares at certain times of the week or seasons of the year. It's almost a western civilization tradition to charge " What the market will bear".

The idea is if a client is wealthy and can afford more, you charge more. He might not be as quick or able to notice any discrepancies or unfair practices (or overbilling). So you go ahead and take advantage of that. As "American" as that practice may seem, it is quite unprofessional and quite unethical. Truly the land surveyors price should be based on what he's going to do. The professional service he will provide should govern price. The status of the client should not govern the price.

But there is another angle to this subject. It does not involve "who" the client is but rather who the "neighbors" are. It does not involve the fee for the professional service but rather the method by which the survey will be done.

I have known of situations where land surveyors were working for a large developer. The development was going into an area where the adjoining neighbors were generally long time folk who did not have a good handle on some of the basics of land ownership. These were people who were perhaps not well educated for what ever reasons. In these cases the survey project would be done in a rather sloppy manner. It was felt that the risk was worth taking. None of these adjoiners would be smart enough or wealthy enough to complain or raise an issue.

I remember a particular situation in northern New Mexico. I had the opportunity to live in New Mexico for almost nine years and I did many surveys for the Federal Government. I also did a few for my own business. In a couple of cases I did some freebies for friends. In this situation a surveying firm went in and did just a horribly sloppy job of surveying out a boundary. They should have run control between two brass caps that were about a half a mile apart and proportioned out some positions along the way to carve up the land in the proper legal method. Rather they went simply record distance from one of the caps and laid out a parcel. This parcel overlapped into the property of a friend of mine by about ten feet.

Neither of the parcels had ever been surveyed before; yet there was a fence over seventy years old between the two parties. This surveying firm had the gall to run out a basically illegal survey and tell their client that their adjoiner was trespassing by ten feet. They suggested that they should make my friend go move the fence. My friend (the adjoiner) contacted me and asked me for my input.

I was absolutely dumbfounded at the approach they took with this survey. They had attempted to bluff everyone in the area with their bogus survey. They even had the nerve to allow this survey of theirs (and accompanying plat) to be entered into a quiet title suit against my friend to try to make him move that fence ten feet. It was a classic case of doing a crummy survey because they didn't think anyone next to them would be smart enough to take care of it.

In a few of the western states, most notably California, the surveying profession has taken on a totally new approach to this question of surveying methodology based on who the adjoiners are. In particular I refer to the public land survey system and the belief held by some registrants that the public land system no longer applies, and you can basically do whatever you want. I've heard that expounded to me on a number of occasions when I was doing public land systems seminars in the west. These people basically feel that a call for a parcel of land by an aliquot part description was nothing more than a metes and bounds based on 1320's, 660's, and ninety degree angles.

I seriously question this reasoning because I don't believe that it is up to the surveyor to make a decision that people's property rights have changed or have moved. The surveyor is to report the facts. It is not his decision to say, "well this legal description system really means this one and so we're just going to do what we want".

My counter argument to this line of reasoning has always been this: What would you do if you were to subdivide a section for a private land owner who had a quarter section of land? Would you do it any differently if there was Federal land in that section? Almost always the surveyor will respond, "well of course, I have to obey the BLM Manual and the legal principles involved because there's Federal land involved". We need to stop and think about this line of reasoning.

How is it that you would survey a piece of private land one way today and would survey it differently tomorrow if the Federal Government owned land there? How is it that land boundaries are changing between private land owners simply based on who else may be in the section? I find this to be an amazing ethical issue and yet some of my fellow registrants in California continue to do some extremely poor and blatantly illegal section subdivisions.

It's time to wake up! It's time to realize that you cannot change the method of survey based on who the adjoiners are. You should not change the price based on who the client is. There is no choice about how to do the survey regardless of who the land owners are. Remember, the adjoiners can change on a daily basis and the interests can come and go. But the most important thing the land surveyor can do, is identify, in the legal and proper process, where peoples' boundaries are located. Those boundaries should never move nor be somehow adjudicated by a land surveyor simply based on who are the neighbors.

Surveying is surveying. Whether you work for the U.S. Government, your own business, or a large civil/survey firm, you must do the job the same way. No matter who lives in the area, or who owns land nearby, the survey should always be done in the same manner.

Chapter 23

EVIDENCE DESTRUCTION

> "Professionals are often delegated an exclusive franchise for the purpose of protecting the public from the unqualfied. In exchange for this exclusive franchise the professional does have moral obligations to the public; this individual is in a position of trust."
>
> *Walt Robillard*

I think it would be hard to find a professional surveyor who believes that it is acceptable to destroy survey evidence. When I refer to survey evidence I'm speaking of corner and monument evidence as well as record evidence. It would be ludicrous for a surveyor to endorse such a policy.

This is not to say that I am not aware of a few situations where professional surveyors have destroyed evidence. I know of a couple examples where a licensed land surveyor destroyed original survey evidence because it did not fit a survey that he had already done and was now trying to defend. I even know of three examples where federal government surveyors have destroyed survey evidence because it did not fit their survey. Quite frankly, such actions are deserving of removal from the job and being stripped of your survey license forever. (In case you haven't noticed, I feel strongly about this issue.)

But I wonder if we have considered the other ways, the subtle ways, that we surveyors can destroy evidence. Even the most conscientious surveyor could destroy survey evidence. This happens in two ways.

First; a surveyor who finds old deteriorating survey evidence on the ground but does not preserve that position by a remonumentation. He has essentially allowed that evidence to further deteriorate and disappear from view. This would be a foolish action to take and yet many surveyors do just that. "There's not enough money in the project to go out and remonument all those corners." In reality the cost of remonumentation of old and obliterated corners should be a basic part of our fee structure and it should be built into every job.

The second way a professional surveyor may actually destroy evidence is by not actually getting the results and findings of his survey into a public record. I applaud the states that have mandatory plat recording laws. Most of those laws require you to file a plat under certain circumstances. The ideal plat law would simply read, "If you set a monument or if you find a monument in the process of a survey, you must file a plat."

It doesn't bother me that it drives up the price of the survey. It is part of the most fundamental cost of doing a survey.

I realize there are many states that do not have mandatory plat recording laws. Others have laws that are so weak and unenforceable that everyone is ignoring them. The truly ethical and professional surveyor is not going to take advantage of that shortcoming in the law. You and I destroy survey evidence, we rob the public of protection of it's rights, we cheat our client, his heirs and assigns, simply by not getting information into the public record.

I've heard dozens of excuses as to why surveyors don't record their plats. Some of them feel it's their clients "private property." That argument insults the entire profession. The reason we are licensed is to "protect the public." The only way to protect the public is by making a public record of the survey. Others have complained that it costs too much. Yet those same people will be arguing in the bar at the annual surveyors convention that surveyors don't make enough money.

As mentioned above, the cost of a plat and the cost of recording that plat as well as the cost of remonumentation of found monuments in the ground should be a fundamental expense that is always covered in the survey. Besides, if you really want another surveyor to accept your work, make it "plain as day" what it is that you did and what you left there on the ground. I personally distrust the surveyor who is constantly hiding information or withholds critical information from me or the public. I don't trust anything those kind of people do.

So there's the challenge my friends. Not one of us would endorse the destruction of record or field evidence. And yet through clever arguments and justifications (and flat out ignorance for the entire purpose of licensing our profession) many surveyors do in fact destroy survey evidence. I encourage you to rise above the norm and lift up our entire profession. This is why we exist.

Chapter 24

THE IMAGE
OF THE SURVEYOR

"All of the technical knowledge in the world is of little value unless a person can also convey this knowledge to others in a convincing manner."

Curtis M. Brown

You've heard it many times. You've discussed and argued it many times. It's the subject of our professional image: How we come across to other people.

Let's not play games with this; our image is defined by many different things. Whether we like it or not, people do judge us and our profession by various things that may or may not have anything to do with our professionalism. The true professional (and for that matter just the intelligent person) will recognize these facts and will at least in some way compensate for them. Even if they disagree with the principle at hand, we will adjust.

For instance, one of the more classic examples of that is dress. The standard survey crew in the field is not well known for setting fashion standards. Let's focus on this example. What if you are the field surveyor but you are meeting with a clients' representative in the office? If you've been in the field all day, to come in dressed that way, would not be very appropriate. Yes, some people are going to argue that no one should judge us on how we look. That's the way our profession is and they ought to just take it the way it is. Yet the fact is that you leave an impression.

What if you were to show up in that other persons' office dressed like that, tracking mud in on the floor? You not only leave a bad impression, but you also impact the bottom line of surveying, the profit margin. You could loose a client. Can you imagine loosing a client who over the lifetime of your business could provide hundreds of thousands of dol-

71

lars of work? Could you imagine loosing just one of those kind of people simply because your crew or your people didn't want to dress properly? That's an example of going ahead and working around one of your personal biases to try to make room for other people's biases.

The whole subject of image goes far beyond just the way we dress. Let's consider some of the other areas. The way we speak; our language. Did you ever consider how that reflects on you? The use of profanity and dirty jokes is a good example. Or perhaps the quality of your vocabulary. I'm not just talking about bowling people over with tons of survey jargon (and there's plenty of it.) But rather the use of intelligent vocabulary and the ability to understand what the client says. All of this reflects on the image.

How about the appearance of your truck? I know that survey trucks are notoriously the most filthy trucks in the world (other than some construction laborers). But what about the general cleanliness of the truck? Or more importantly, its' state of repair or disrepair? This can reflect a lot on the company's' policies and attitudes as seen through the eyes of the client or potential client. Conversely, if all your crews are working out of Cadillacs, it's going to send a different wrong message to those potential clients.

What about the way your office looks? What is the "presence" it conveys when someone comes in the front door? Is it neat and clean or is it cluttered? Does it have the appearance of organization, competence and professionalism? Or does it give way to the sense of total confusion and chaos? I know the back room of many a survey office basically operates on chaos. But the reception area and boss's desk need to be something different.

You might wonder how this might fit in with ethics. The professional land surveyor will be keenly aware of his ethical responsibilities to his client, to society, but also to his own staff and his own profession. These sort of things, the whole area of image, reflect on many more people than these residing in your office. It reflects on the profession as a whole. For all of us to continue to succeed and improve in our professional standing in the world and amongst ourselves, we need to be contributing to one another in this very important arena.

Chapter 25

THE LOW BID

"It is wise to keep in mind that no success is necessarily final."
Unknown

Having personally worked in both the private sector, in business for myself and for the Government, I recognize the reality of the processes needed to obtain survey work. The basic methods by which land surveyors obtain work from a client are:

Referrals: Either the client is referred to us by another client, by another land surveyor or by a disinterested third party such as a title company or engineering firm.

Advertising: This would include ads in professional journals, newspapers and especially the yellow pages.

Referrals from professional associations: This includes the surveying associations in some states or other professional registrant organizations which include referrals as part of their service.

Self Marketing: The surveyor who simply markets himself and his skills through civic organizations, community involvement or otherwise is letting the world know he exists through means other than advertising.

A professional qualifications process: This would include most projects coming from government of all levels where the selection is made based on the qualifications of the firm.

Each of these methods involve different dynamics, efforts and approaches. I personally have deeply appreciated the change for the Federal Government from a low bid process for surveying contracts to the "Brooks Bill" procedures which is a qualifications based system. Truly, the quality of work has improved. Granted, at times, the selection process by any particular government entity may not be done in a completely fair and unbiased manner, but the work seems to be well spread out. The award of the work to the surveyor is based on something other than his price.

But with the remaining four methods of acquiring work, the surveyor is often dealing with a price bidding situation. There is a real danger to the individual surveyor on any given project as well as to our profession as a whole when we get into the low bid arena. To the surveyor on a project the low bid may just be a disastrous financial loss before completion of the project. For the profession as a whole it creates a very unprofessional image of surveyors cutting each others throats to get some work.

I had may own surveying business for a short time in the Flagstaff, Arizona area. In almost every project I ever did I was in the situation of having to bid for the work. I know that feeling of not having a whole lot of work on the books. Yet you have employees whose livelihoods and benefits depend upon you. I know that gut-wrenching feeling when you know that you've got to lay off employees too. So I can really understand that desire and the tendency of a land surveyor to go ahead and bid a job at a lower price then he normally would. Sometimes we are ready to "Just pay the wages". Truly it is a noble concept that one would be willing to forgo profit in order to hold on to valued employees.

But what is the net effect? Unfortunately surveyors in a given geographical area will continue to under cut each other. It continues until every job you are doing, (at least of a certain type) is down to a price that not only lacks a profit, but probably doesn't even pay the cost. In fact, surveyors are infamous for this practice throughout the United States.

Another great disadvantage that comes along with the low bid survey is the tendency to short cut a job in order to somehow offset the fact that you don't have enough money in it to cover your costs. Those short cuts can come in many different ways. In particular they show up in a failure to research the record accurately, or a failure to search for corner evidence on the ground for a sufficient amount of time, or a failure to do a complete job. Surveyors are willing to take great risks and incur great liability to try to break even on a project.

I know surveyors get frustrated at times when people compare us to other professions, but I cannot resist the temptation. Let's think for a moment about the medical profession. Aren't you glad that those people are not low bidding on your surgery that you'll need next year? It's comforting to know that. We can at least have a higher level of confidence that things are going to be done right. If they aren't done right then the doctor is just a crook or a quack. But if they were bidding open heart surgery like surveyors bid land surveys we would have this major medical operation down to a few hundred dollars. Can you imagine the quality of care you'd get? Can you imagine how much time the doctor would really have to diagnose your problem let alone perform the surgery? It's a frightening thought. The doctors would never do such a thing. They would never take aim at their own profession and their own livelihood with a shotgun and absolutely decimate it. Yet we surveyors do just that!

One of the basic ways in which a surveyor gets work is through their ad in the yellow pages. I have often referred to this process in my seminars as "Yellow Page Roulette". Similar to "Russian Roulette" where the participant takes several major risks to see if he lives or not, land owners and developers and construction companies will get on the phone and go through the yellow pages and call various surveying firms in the area asking for a quote.

I can understand providing a quote for things similar to construction staking where progress and production are very predictable. But what about the land survey? There is little that I have ever experienced that is predictable in the boundary surveying business. The only thing in a boundary survey that I could safely predict (plus or minus ten percent of my time) would be the actual field traverse computations and drawing of a plat. Those are the simple, almost nonprofessional duties that one would perform in a boundary survey.

Land surveys are more affected by the quality and caliber of records research and corner evidence search and analyses than any other factor. And these two factors are the most unpredictable things in the profession. It boggles my mind to consider some of the prices that I threw out over the phone to do a survey. Thankfully, I didn't get some of those jobs. But unfortunately, I did get some of the others. I found myself facing many times, perhaps the most difficult ethical questions I've ever had to face: those involving what I was going to do with a project in which I didn't have enough money.

I can see exceptions to this situation. There were opportunities in Flagstaff to propose on a project that was next door to (or integrally involved with) a project I had done in the past. Obviously the surveying and engineering firm that has been in an area for a long time and has a vast set of records has a tremendous advantage over firms that have recently moved into the area. New registrants in an area face similar hurdles. This is certainly a valid means of obtaining work and it is also a valid warning to the new registrant and the firm that opens up a new branch office in a community in which they have very little background. There is significant danger in that kind of effort.

The profession as a whole has suffered from the low bid approach to work. Consider the prices that we're getting in some parts of the country for certain professional survey work. In many areas the mortgage inspection survey or (improvement location certificate) is down under one hundred dollars. Many surveyors are providing surveys for large tracts of rural land that require significant survey control work. They have allowed the value of the land to influence the value of the survey. There certainly would be cases where I would advise a potential client that "this survey will cost far more than your land is worth, I don't think you really need to do this." Surveyors hate to throw away work and yet I would much rather take that approach then to go out and flash in a survey or loose money. I'm in the surveying business to make money.

I think that one of the reasons surveyors are so easily sucked into the "black hole" of low bid work is that surveyors as a group are not very good businessmen. I can say that about myself and the vast majority of my fellow professionals. Many of us are excellent technical and legal experts and yet when it comes to making a dollar we're not very good at it. Worse yet, even when we make a dollar we are not very good at managing it and understanding what it took to earn that dollar.

The average surveying firm is a small almost "mom and pop organization." I personally know dozens of surveyors in many states who are no longer in business. Many have even changed professions simply because they couldn't make a living at surveying.

Let's do ourselves a favor. Let's do our entire profession a favor. Let's raise ourselves up from the basement of the "low bid mentality." Let's look for alternate ways of proposing on a job. These methods could include:

- Ranges of prices that are quoted.
- Estimates with a number of caveats well explained to the client.
- We could also try to sell ourselves and our expertise to the client truly based on our professional skills and abilities and experience.

We must stay away from the price arena. I personally have found this latter method to work quite well. Many times I used my other skills, such as presentation skills, ability to organize and present materials and just a basic honest approach to things. I often won over people to give me the job regardless of the price. That was the only thing that kept my business going. Eventually my business went under. I could no longer afford to run a surveying business there in Flagstaff. There was stiff competition and they were good surveyors. The main reason that I got out of the surveying business was the majority of my work was contracting from the Federal Government. At that time it was all low bid and I went broke.

The low bid is essentially an unethical practice. I recognize the need and the reality; the need to operate in that forum and the reality of how business works. But my friends, surely there are better ways for us to conduct ourselves on the whole. Look for ways to market yourself. Line up clients through means other than a low-ball no-profit price.

Chapter 26

CONES

"It is unprofessional to act in any manner or to engage in any practice that will tend to bring discredit on the honor or dignity of the surveying profession."

Curtis M. Brown

In 1976 I was employed by a large consulting civil survey firm in Phoenix, Arizona. We did a tremendous amount of work around the Phoenix area. If you've ever worked in Phoenix or in similar cities you know that almost every project involves you setting up in the middle of the streets. In Phoenix those were very busy streets. Safety was a concern and it was customary to have fluorescent traffic cones on the road. They would lead up to you and your other equipment to warn drivers and to protect you and your crew.

As time goes on cones get damaged, disappear or simply wear out. We were just starting a new project on some busy streets and I knew we would need some additional cones. I went to the chief engineer of the firm who was overseeing the survey department. I told him that I needed about ten new cones. He looked me straight in the eye and told me, "Next time you go by an Arizona Department of Transportation (ADOT) project just stop and pick up some of their cones." I assumed he was joking and I laughed. But it was obvious that he was dead serious. He had no intentions of spending a few bucks to provide for the safety of my crew. He wanted me to go out and steal traffic cones from the State Government.

I had no intentions of stealing cones and I was really offended by the unethical behavior of this man. He was insisting that I steal something if I wanted to provide safety for my crew. There's no way I could go with that solution and yet he controlled all of the purse strings. What could you do?

Ethics often comes down to basic honesty. I cannot claim perfection in this area over the past 40 years. But this case angered me because it was tied to our safety. I was also disturbed that this large firm would stoop so low to procure such a small ticket item. At the time they were fat and happy with lots of work. Why should they risk a stupid stunt like this to save a hundred dollars? And what would have happened to me personally if I'd been caught taking this clandestine procurement action? In retrospect, I was dealing with a cone-head of a supervisor.

I found a much better way to do it. I went and "borrowed" some cones from one of the other crews in the same firm. I assume they had to go to the ADOT project to get their cones! But whatever happened to the slogan, "Safety First"?

Chapter 27

MURPHY USED TO BE
A SURVEYOR

> "If I had only known, I would have been a locksmith."
> *Albert Einstein*

(This chapter is co-authored by my very good friend, associate, and fellow professional, Wyman Bontrager.)

    In many a conversation with a wide variety of people, you are asked what you do for a living. Once I have announced I'm a surveyor, they will often respond, "Oh, I used to survey". This of course is a reference to their 3 months experience holding a rod or the "dumb end" of the tape. But of course, they assume that vast amount of experience gave them all knowledge of being a professional surveyor. I have come to discover that a very famous person that we all know well also used to be a surveyor!

    Poor Murphy, struck by the misfortunes in life and depressed by what seemed to be the natural laws enforced against him, picked up a chain and started "Murphy's Location Service, Limited". Their motto on the survey truck door was "You tell us where you want it and we'll prove it's there". Later in life when he discovered that publishing was the better part of valor (and much better income), Murphy remembered his wasted past, surveying where no sane man had gone before. Here are Murphy's laws discovered while sighting the lines of geodetic uncertainty.

Murphy's Laws of Surveying:

1. If the client's deed matches the location of possession, all adjoiners hold conflicting senior deeds.

2. The battery life of the total station is proportional to the distance traveled from the replacement battery.

3. When a true line is to be run, the single tree in the meadow is on line.

4. If a monument is set in easily dug ground, it is in the wrong place. (Corollary: If a monument is poorly set, it is in the right place.)

5. If a point for an angle point falls exactly on a previously set junior monument, both surveyors have made the same error.

6. The chance for cloudy skies are proportional to the necessity of the astronomic observation.

7. Radio failure is proportional to the distance and noise level occurring in the interval between a separated survey party.

8. A data collector used by a new instrument person will only collect rejected measurements.

9. The center of a section is a myth.

10. The original monuments found by the B.L.M. (Bureau of Land Management) spent the intervening years moving from one fence corner to the other.

11. All lost monuments reappear within three years of a newly set replacement monument.

12. All engineers used to survey.

13. True meridian north is a matter of opinion.

14. The only tree trimmed or removed for surveying purposes is someone's favorite tree.

15. All adult persons encountered during the course of a survey have been employed on survey crews before getting honest work.

16. An error in platting is always discovered within the second week after recordation of survey.

17. Dead and down Bearing Trees always fall with the blaze on the down side.

18. Calibration for metal chains and electronic distance meters use different definitions of a standard foot.

19. Plumb is in the eyes of the beholder.

20 The adjoining landowner will only be present when the survey crew is locating his encroaching improvements.

21. A monument to locate the center of the section breeds providing a minimum of three offspring within a 30 foot radius.

22. Most misclosures of survey are caused by the imperceptible movement of ground reflecting the gravitational force of the sun and moon as the earth rotates on its axis.

23. All bearings run in two directions, i.e., "North" and "The other North!"

24. Your latest and greatest new model of total station will be dropped from the manufacturer's line, one week after purchase, due to a revolutionary development in the field of electronics and listed at one half the cost.

25. ACSM conventions are held 1000 plus miles from home except the year you win the big subdivision construction contract.

26. The seller will rebuild his fence the day after you finished the location certificate.

27. All errors of survey will be found upon review of the field book, two weeks after all monuments have been set.

28. If the profit on the contract is enough to live on, your client has moved to Mexico the day after the survey was finished to avoid the embezzlement charges.

29. A section that measures one square mile is a mistake.

30. The area of an aliquot part is proportional to the financial and political condition of the owner. (This is also known as the double-proportionate method.)

31. If only one stone corner can be found in the section, it is from the 1883 suspended survey.

32. Vertical angles steadily decrease after two in the afternoon, as does production.

33. Competitive bidding for contracts requires the adoption of a new life style, i.e. you will no longer eat, wear clothes, or see a roof over your head.

34. Tripod legs can only be kicked after the third recorded angle.

35. If property lines are in doubt, the best evidence is the client's memory.

36. If a fence is built approximately along the boundary of private and government lands, it will always include a minimum offset of 5 feet into government lands.

37. The centerline easement surveyed for road construction will provide hours of laughs to the design engineer.

38. Wind velocity is proportional to the importance of the shot and the unsteadiness of the tripod.

39. Winters exist to give surveyors time to prepare a court defense for the past field season.

40. The boundary agreement, to which your client referred, will have only one party.

41. The software manual you received with your COGO program Ver.2.5 will be for Ver.2.4.

42. All points set for the construction survey of the new interstate will be relocated by Earth First, mid-way through the project.

43. The per cent chance of damage to your new GPS receiver is determined by division of the purchase price by the purchase price multiplied by 100.

44. A bulldozer driven within a 100 foot radius of a monument will always remove the monument.

45. Surveyors have been the only scientists to prove that water flows uphill.

46. The State Board of Engineers and Surveyors is compromised of rail road engineers whose distant relatives knew at least one engineer or surveyor.

47. The last page of the field book that proves the question before the court will be the only page affected by last night's coffee spill.

48. If your client's deed calls for the line to be coincident with the adjoiners, the adjoining deed will call your client's deed as non-coincident.

49. State plane coordinates are based on either lame or murky conical projections.

50. The higher the elevation of a project, the later in the fall it will be scheduled.

51. A GPS unit left unattended will be adversely possessed.

52. True understanding of the legal principles of surveying are understood only by the presiding judge.

53. All states acknowledge reciprocal licensing except the one in which you want to practice.

54. All LSIT's think they are surveyors.

55. The necessary items needed on your plat will be 5% larger then the space allotted.

56. A smiling, willing to help, government official does not understand what you want.

57. If your boss is smiling and willing to help, see #56.

58. A lath without a knot does not exist.

59. The distance capability as advertised for your EDM will be 10 feet shorter than your shot.

60. Hiring your brother-in-law as rodman will be as efficient as using a plumb bob with baling wire for string.

61. The magnitude of the error in your corner placement is directly proportional to the amount of fluorescent pink flagging you have tied around the area.

62. A found monument is not officially accepted until you tie your flagging around it. (This is part of the ceremonial procedure of accepting evidence).

63. The highest and most difficult to reach limb is the only thing blocking the EDM shot.

64. The line you decide to "just cut everything" will be the wrong line.

65. Closures of 1 part in 100,000 or more are the result of incredibly compensating errors.

We welcome your additional laws, as we've all experienced the realities of working in our profession. Above all, don't let old Murphy get you down! Surveying is still the greatest profession on Earth.......as long as you don't set any monuments or file any plats.

Now you may be wondering, "what does this have to do with ethics?" The answer is: Ethical issues are just as common and just as sneaky as old Murphy and his laws. So keep an eye out for both in your daily practice as a professional land surveyor.

And be sure to have, and keep, a sense of humor.

Chapter 28

TWO PLUS TWO

"You will never be able to truly step inside another person, to see the world as he sees it, until you develop the pure desire, the strength of character,...., as well as the empathic listening skills to do it."

Stephen R. Covey

In the seminars that I've done around the country for the last seven years, I've probably told one particular joke about a couple hundred times. It really speaks the truth about surveyors and our relationship with the other professions we deal with. That joke has been modified over the years. I honestly do not remember where I first heard it nor who it was that initially told it to me. Therefore I cannot claim full responsibility for it. However, I have at least doubled it's content.

I'd like to share that story with you just to make the point one more time.

It seems there was considerable debate amongst the professions as to the answer to a simple mathematical question.

What is 2 plus 2?

The Government was very concerned about this breakdown in the basic fabric of our society. So they proposed a million dollar grant be given to the National Science Foundation to study this problem. They had to determine the true exact answer to the mathematical equation 2 plus 2. After the foundation put together a plan for the study and reviewed all of the available publications and documents on this and related subjects, it was decided to do interviews with representatives of several different professions to get their opinions.

The first person they brought in was a registered Civil Engineer. The interviewer posed the question right off. "Sir you're a Civil Engineer. In your expert technical opinion, what is 2 plus 2?"

The Civil Engineer responded, "It's 4.00."

The next person brought in was a licensed Land Surveyor. Surveyors were known for their precision and accuracy, they were great mathematicians, or so the foundation had heard. They posed this question to the Land Surveyor, "What is 2 plus 2?"

The surveyor responded, "Well, it's four......more or less". [1]

The next person they brought in was a computer whiz expertly trained in Autocad. The cad operator was asked, "What is 2 plus 2?"

He responded, "It's 4.0000000."

The next person they brought in was a Title Officer. They asked him, "What is 2 plus 2?"

He responded very quickly, "I don't know."

They said, "Don't you care what 2 plus 2 is?"

"No, I don't care!

"What are you going to do about it?"

"The only thing I'm going to do about it is exclude it on Schedule B. We don't care what 2 plus 2 is."

A real estate appraiser was brought in and asked, "What is 2 plus 2?"

He responded, "Well, based on other 2 plus 2's in the area and based on some 2 plus 3's and some 2 plus 1's and square roots of 2 and some other similar things in the area, after having made adjustments for time and river frontage and financing I have determined that 2 plus 2 is 4. However that opinion is only good for six months."

A real estate salesman was brought in and asked, "What is 2 plus 2?" He immediately left the room and went back out to his bright red Cadillac parked in the parking lot and retrieved out of his trunk a copy of the real estate primer book. He brought it back in and spent about 20 minutes looking in it, researching through it. Finally, he looked up and sheepishly asked the interviewer, "It's four, right?"

It was obvious to the interviewer that there was a great deal of confusion between these professions. It was decided to get a legal opinion. They brought in an attorney and asked him straight out, "What is 2 plus 2?"

The attorney looked back and smiled as he said, "What would you like it to be?"

Let us never forget the many different professions we deal with and how they all view the world. They see it through very different eyes. The true professional surveyor understands that. The truly ethical surveyor never takes advantage of it.

---

[1] For those of you that wear both the engineers and surveyors hat, it is hoped you know when to change hats!

Chapter 29

THE COUNTY SURVEYOR

> "Nearly all men can stand adversity, but if you want to test a man's character, give him power."
>
> *Abraham Lincoln*

Most states still have the office of "County Surveyor" on their books. In many locations this is an elected position. In others it is just another civil service position. In either case, many counties (or parishes) have a "hit and miss" approach as to whether the job is filled and with what type of individual. In several states, the office does not even require licensure. Needless to say, there are some interesting ethical issues out there for those who hold this title.

I have met well over a hundred county surveyors, primarily through the seminar process. These people are for the most part very conscientious surveyors, who take their job and title very seriously. I salute you folks and thank you for your efforts and in many cases, your leadership in our profession. The original concept of a county surveyor was one of leadership and stewardship. That office was to be the technical expert for the growing and developing counties established in the frontier. It was an important office, and still is.

In light of the diverse applications of the title, I recognize some of my comments may not be relevant to all who read this chapter. Please bear with me, and realize that the principles go far beyond just the County Surveyor. Any surveyor in some sort of position that allows for review of the work of others can gain from this discussion.

Across the country I've heard two main complaints against the office of County Surveyor. These complaints highlight the ethical issues this group of individuals should consider. These two concerns are:

- Those who are elected are not always well qualified.
- Those in a plat review role are often not qualified, fair, or reasonable.

Let's look at these in a little more detail.

The election process is one which certainly leaves room for doubt as to whether the right person got the job. Look at the national and state elections. Often the biggest spender, biggest mouth, or biggest liar actually wins instead of the best qualified person. This author has often wondered why we elect many of our positions that actually require a technical skill as opposed to a political skill. Assessors, sheriffs, treasurers, and similar offices surely would be better filled through regular civil service functions than by election.

I remember when I worked in central Missouri that our competitor in town always got one of his chainmen to run for County Surveyor. This fellow was not licensed, and was not often sober. But he won the election being the only person on the ballot. His boss would reward him with a $50 bonus for winning, which was probably guzzled within 24 hours of the election victory. This seemed to provide the employer the opportunity (and right as he saw it) to take the County records home. It was standard procedure to find out at the courthouse that the records we needed had been taken by the County Surveyor. This was an insult, a joke, and an illegal unethical act.

If you are in an elected position as a County Surveyor and are not licensed, you still have a legal and ethical responsibility. Frankly, states with this inane mixture of rules should be lobbied heavily to change the rule. The most basic rule would be the requirement to be licensed! Without the license the position is a sham and a waste of taxpayer dollars. If the incumbent is doing good work in filing, records management, and research assistance, this is a great technical service. My point is that the non-licensed County Surveyor has *absolutely no business* making any professional decision or judgments. And he or she should respect the *sanctity* of the public records. I don't care what some asinine state law says....it isn't right.

Whether a County Surveyor is licensed or not, one of the greatest criticisms of the position is the role of plat review. I know of counties where the Surveyor's staff is huge, exceeding 100 people. Their task is to review and bleed all over any and all plats submitted for county use.

I have no problem with a county official reviewing my plat to be sure it has all the elements required by law (i.e.— a north arrow, scale, my name and seal, etc.). I don't mind an official reviewing my plat and *commenting* on his disagreements with my approach or procedure. In fact, this can be quite healthy. (See the chapter on Peer Review). What I cannot tolerate is an official who reviews my plat, disagrees with my procedure, and *has the power to refuse to record my plat. I object to this because:*

- I am licensed by the state to perform surveys for my client. I should not have to change my professional judgment simply to get my plat into the public record. (This includes the process of the County adding a notation to your plat that says they disagree with your survey...this is intolerable.)

- I have not met too many County Surveyors whose expertise was broad enough and whose experience was sufficient to make such "life and death" judgments on my survey. I can and have made mistakes, but none caught by an official.

In many counties with this official review and judgment power, the plats are returned with redlines all over them, showing adjustments of .01 foot or less to be made to make a perfect closure. Any surveyor worth his/her salt knows this is the least important thing to be corrected. And after all the misclosure comments, plus my north arrow was not large enough, they did not notice I used a completely different piece of evidence for a corner than they used. In other words, they completely missed the most important element of the survey, and spent all their time making it look pretty. Have we lost our way or what?

If you are a County Surveyor with such review power, please consider the ethics of this action. Even if state law gives you this power, are you that well versed? Do you have the ability to review all the survey, including the elements which really make the difference (like the evidence in the field)? If you have a difference of professional opinion with a surveyor, do you really have to exercise your *power* and reject his work? There is tremendous danger with this veto power imbued on some officials. Please take it seriously, but also professionally. Ethically speaking, your office and a whole book of state statutes does not confer on you perfect judgment. We need to be professional about it.

Unfortunately, some with this title have had it go to their head. I know of several instances where the local survey community spends more time kissing the County Surveyor's butt than they do on the survey itself. They just know the keys to getting plats approved is to win friends and influence people, like the County Surveyor.

In one county I know of a County Surveyor who had his license revoked by the Board. He still held the office for a few weeks. He decided it was his prerogative to remove a good percentage of all his plats and notes from the courthouse permanently. He figured they had his number on them, so they were his property. The fact they'd been done under the authority of the office, and with tax dollars did not seem to matter to him. So now I cannot get all kinds of information from the courthouse, because it is sitting in his basement collecting dust. This is a blatant abuse of the office, and one of the more unethical actions I've ever seen an elected surveyor take.

Elected or not, this office carries a heavy responsibility to the public and the profession. Do not abuse that power. Do not allow a conflict of interest to appear. Take your job seriously, ethically, and responsibly.

And for all those of you in this office who are doing a good, fair, and ethical job, I want to say "thank you". Keep up the good work.

Chapter 30

BLACK AND WHITE

"When a man has had a training in one of the exact sciences, where every problem within its purview is supposed to be susceptible of accurate solution, he is likely to be not a little impatient when he is told that, under some circumstances, he must recognize inaccuracies, and govern his action by facts which lead him away from the results which theoretically he ought to reach. Observation warrants us in saying that this remark may frequently be made of surveyors."
Thomas M. Cooley, Chief Justice, Michigan Supreme Court, 1864-1885

As most people who know me would tell you, I'm a big believer in education. Of course I'm a bigger believer in continuing education than I am of anything else. That's because of the condition of our profession. I fully support the four year degree concept in surveying.

What has disturbed me is the trend that I've seen in some four year graduates. Some of them are excellent surveyors, people I'm honored to be associated with or just know. But there is a general trend that I have noticed with a number of them. Some are mathematical wizards but also legal buffoons. (Please don't misunderstand my comments. My point is simply that we have educated them in only one side of surveying and we have left out a very important half.)

I find many a new (or aspiring surveyor) who is far more interested and educated in closures than in proper survey procedure. There is a stronger desire to get a traverse to close one part in one hundred thousand than there is to find the section corner and use it properly.

So over the last several years in doing my seminars I have seen an interesting effect of this situation. My seminars are usually on the subject of the public land survey system. But it doesn't really matter what the subject is, all the principles are the same. I try to cover the fundamental legal principles involved in survey law. But I'm amazed at the response of some of the students. They indicate not only a lack of interest in these principles but an apathy for learning them.

They are shocked when I suggest that precise measurements are of secondary interest to the professional surveyor. They are dumbfounded by my comments that when using bearing trees, if the distances match within a couple links, it's good enough. They think I'm crazy.

All of this comes from a strange phenomenon that is occurring in our profession. That is the desire to have everything in black and white. Mathematics are black and white. Closures are measurable things, precision, statistics, all of these are things which have walls and rules that apply to them. But what about the legal principles involved? There are some cut and dry things, but almost every project is unique. There is always something different to be dealt with. There are professional judgments to be made. I'm not going to apologize that surveying is not a black and white profession. In fact, that is why it is a regulated licensed profession.

So how do you view land surveying? Do you see it as a purely black and white cookbook solution for every problem? I'm afraid that's what many of the people coming into our profession now are looking for. They want to know the cookbook answer to solve the problems and they are not skilled. Neither are they willing to acquire these skills that are required to be able to make the proper legal judgments.

The most critical skills that I am speaking about are:

- The records research and analysis skills
- Corner research and analysis skills.
- Law and legal principles

It is extremely difficult to teach each of these subjects in a school classroom environment. Some things can be covered in the classroom, but I'm afraid there are significant limitations. Perhaps we really do need some form of apprenticeship that goes beyond "How many years experience do you have?" There are some people with one year of experience ten times.

Again, I am not being critical of the universities and colleges who are providing our future surveyors. But I do offer those graduates a challenge: Recognize the limitations your education has built in to it. Then go on to round out that education.

No, the surveying world is not black and white. If that makes you nervous, *find another profession!* It is complex and legalistic. It is, in fact, a function of law. Erase that desire for a black and white cookbook solution. The sooner you do, the sooner you'll step into a much larger role. You'll then convert from licensed measurer....to professional surveyor.

Chapter 31

A FALSE CORNER

" The lip of truth will be established forever; but a lying
tongue is but for a moment.

Proverbs 12:19

In 1980 I was employed by a civil engineering firm in northern Arizona. I was running the survey department which basically ran two crews. It had been a very slow year for surveyors and engineers. The Flagstaff economy was in one of those greatly fluctuating periods. There hadn't been a whole lot of work for three or four months.

One day a woman walked into the office who owned 120 acres of land that was surrounded by National Forest. It was some 30 miles southeast of Flagstaff. The project was going to include subdivisions of two sections. There was a large lake that existed in the area that did not exist at the time of the original survey. The lake was only three or four feet deep at best, but in the original survey notes they said that they set section corners all the way through. So it created a complicated and rather large-scale project to gain control; perhaps across the lake.

My boss (a PE/LS) went out to look at the project. Feeling rather desperate (not only for work but also to justify our existence) he proposed a rather low price for the project. He basically wanted to get the job at our cost with no profit involved. During those times I didn't have too much of a complaint with that. But in spite of his registration status (and his good intentions) I felt he had very little knowledge and background as to what a project like this would take.

The price that we settled on was $2200. I felt our costs would be at least double that. I protested this price but was told to go ahead and do the job. Three of us went to the field. After four days of work we had only turned up two of the 12 controlling corners we needed. One of the critical problems was a missing section corner. If that section corner could be found it would reduce significantly the number of corners we needed. We did some traversing, we tied in fences. We tried to identify as much in the field that we could and get an idea as to where some of those corners might be.

But in the end we could not identify that section corner with any clarity. One of the pieces of information we did gather during the course of our research was an old Bureau of Public Roads plat that showed a tie from a P.C. in a highway over to this section corner. We actually had a bearing and a distance. There were in the area two (and possibly three) locations where the road may have been at that time. We felt somewhat comfortable with an approximate location of that road. We determined where the P.C. would have been, plus or minus 100 feet.

Using that area and projecting the tie over to the section corner, we made an extensive search for that corner. But there was absolutely no evidence of anything ever having been there. Further, there was no evidence in that area that there had ever been any manmade disturbance. In other words it was very suspicious why the corner monument was not there if that was the correct location. It was my feeling that we needed to run some additional control to verify that this was a valid obliterated corner.

My supervisor disagreed. We had already spent about $4000 dollars on the project and we were no where near subdividing the two sections. He told me to go out and simply set that corner off the tie. I told him that as a licensed surveyor I could not do that. I needed to feel more comfortable about it.

It really wasn't my call since in Arizona you have to be an officer of the corporation to use your license on their behalf. I was not an officer of the corporation. So it was his laxness and his number that would go on it. I was hoping that he would listen to reason.

The following Monday morning I came into work and found a plat fully drawn sitting on the drafting table with his seal and signature. It was a plat of this survey. I was amazed to see that the survey had been completed and upon inspection of the plat, I was even more amazed to see that he found that section corner. He claimed to have found a marked stone in a large mound of stones. I knew for certain that it did not exist! So I got in the truck and drove back out to the project site. Sure enough, there right in the middle of the area that we had searched, was a marked stone in a mound of stones. He and a good friend of his (who was another LS), went out on the weekend and created a false corner. They then proceeded to use it to subdivide the two sections. Then they produced a plat, signed it, and sent it out of there before there was any input from me.

It does not bother me to accept obliterated evidence if it is as the BLM Manual describes in section 5-9, if the position is validated by ties to topographic features and other factors. I may have even accepted that position as an obliterated corner, but what bothered me here was that he had not claimed that he set it as an obliterated corner. Rather, he claimed he found an original corner. He made an original corner himself that weekend.

We had had a few discussions before, but that afternoon we had one big discussion on ethical issues. His entire justification was that we had already spent too much money and had to finish the job. My point was that I was glad it was his LS number on there and that I wouldn't dare do something that illegal and that unethical or improper. He filed his plat. He sent her a bill. She paid her $2200 for absolutely nothing; an absolutely worthless survey.

Within a week he and I had several continued discussions on the ethics of this and other actions. He made it clear to me that he was a PE and an LS and that his judgment was far better than mine. He was going to continue to tell me exactly what to do. He made it clear that I was not going to overrun the cost on any jobs ever again.

Friday was my last day. I resigned.

Chapter 32

THE LIBRARY

"Education's purpose is to replace an empty mind with an open one."
Malcolm S. Forbes

When I first started surveying there was a small collection of surveying books one could choose from for which to study. A fellow named Dave Slagle was my boss in 1974. He gave me a copy of the 1947 BLM Manual (*Manual of Instructions for Survey of the Public Lands of the United States*) . I cherished that book, as it was the first survey book I ever looked at. I did not understand much of it, but it was an important beginning for me.

In the ensuing years dozens more books have been produced, including this one you're reading! Some of the books are really useful, and some rather mediocre. But there is a tremendous choice. Whatever subject area you need more information on, it is available to you. Perhaps you need technical information on legal land descriptions, or public lands issues, or water boundaries. Maybe you need help in managing your business, dealing with personnel issues, or in technical writing. Or maybe you really enjoy survey history and the instruments involved. No matter the need or interest there is a resource ready for you.

What kind of professional library do you have? Does it have a wide variety of subjects and viewpoints? Do you encourage your employees to browse your library? Have you ever given a book to an "up and coming" employee? What a wonderful way to strengthen your own profession!

I know some surveyors who don't want their employees to read or be educated; they fear the employee knowing more than they do! Then they might have to pay them more. On the contrary, the more your employees know, the better off you'll be. What a great year-end bonus, or longevity award....give them a book about their profession.

Unread books in a library are a crime. Set the example for your employees. Let them see you using the books, referencing them, underlining words in their text. We all experience having an employee in the office on a cold snowy day and no particular work for them to be doing. Hand them a good surveying book and tell them to read a couple chapters, on company time! (You may need to keep smelling salts on hand for the first few times you do this). This is a great way to "examine" your employees. You will learn who has the aptitude and attitude to be a good surveyor. You will find those that really can focus in on certain types of work, and which subject areas are not their "cup of tea".

Education has never been a waste of time. Invest in your profession, and in yourself. Maintain a good solid professional library....and spread the word!

Chapter 33

CREW SAFETY

> ""Employers must provide a workplace free from recognized hazards that are causing or are likely to cause death or serious physical hardship to employees."
>
> *OSHA Act, 1970*

An important ethical issue is that of the safety of your crew. I've seen some extremely dangerous things done with crews to save a few bucks. (See the Chapter on "Cones".) Is there an ethical issue here?

Your crew is your responsibility. They should be equipped with the basic safety gear for the project at hand. It does not have to be the latest and greatest, but it needs to be adequate. Over the course of a year, the average survey crew encounters hundreds of hazardous situations. It would pay for management to sit down with the field crews and review the situations, potentials, and solutions. Such "brainstorming" could help prevent accidents, lawsuits, and even the death of your employees.

Let's look at some possible situations:

- Traffic safety. Cones, flags, signs, vests, and lights on the truck are important, especially in highway or heavy city traffic areas.

- Construction sites. Hard hats, vests, even eye and ear protection are issues worth exploring.

- Rural field situations. Again, vests (especially during hunting season), hard hats, chain saw chaps, gloves, snake bite kits, goggles, first aid kits, and some minor first aid training are basic items one should consider.

- HAZMAT. The first half of the 20th century has provided many hazardous material sites for us to work in and clean up in the last half of the century. Be aware of the equipment, cautions, and training your people need in these very dangerous and often complex sites. If you know a site has asbestos or high levels of radon, for instance, are you being fully open to your employees about this situation? Do they have a choice about working in a HAZMAT site?

- Your vehicle. The truck, van, or auto your crews may work out of should be in good operating condition. Safety features should be working. If you are using specialized tools or accessories, the crew should be trained in their use.

I recognize these are some basic safety issues. You've probably thought them through already. But be sure your crews themselves "think safety". When they run out of things like cones, they should feel free to ask for more or replenish the supply themselves. The real question: Do you have a safety atmosphere in your firm? It is an ethical, moral, and legal issue. It will save you from legal action. And it will save your crew. Think safety, and think ethics—they go hand in hand.

If you are employed by a firm which has a lax or non-existent attitude toward safety, you have three options.

- Suggest they improve their safety preparedness. If this upsets them, go to #2 below.

- Quit.

- Wait till you get hurt and then sue their pants off. They deserve it.

In reality, it's your life or health they are taking for granted. If they are too stupid to see safety as an important issue, you should move on anyway. That sort of firm is a lousy place to work. A survey business with a poor attitude toward safety simply has their ethical fly open. Drop them quick!

Chapter 34

IT ISN'T THERE

> "Proportioning is admission that we have failed as a profession."
>
> *Ken Witt, BLM, retired*

In discussions with many surveyors over the years and in many locations, I have discovered a very dangerous attitude that exists in some members of our profession. It is a belief that the original evidence of previous surveys isn't there. They believe that the original surveys were paper jobs done in a bar room.

The danger of such an attitude is simple. It causes the person saying it or believing it to not look as hard. They won't spend those few extra minutes researching the record. They won't go out with the needed positive attitude; an attitude that wants to look for what's on the ground; that's willing to spend a few extra minutes to look for that evidence; that's willing to use some imagination to solve a puzzle.

My personal experience is that one often finds a well set monument in the field which is a few feet away from the true corner position, as evidenced on the ground. Too many surveyors walk away too quickly from a search area. Is there any more basic function of the surveyor than to find the true corner point? How can you afford to hire people with attitudes which do not lend themselves to this basic function of discovery?

If you have employees who work in the field with that kind of an attitude you need to get them out of the land surveying business. Have them stake bridges. That attitude is just too dangerous. Those in the Midwest and eastern parts of the country certainly have a right to be a little more negative about the possibilities of finding original evidence. You might be surprised to know though that many people even in the western states where the original surveys are perhaps only 80 to 100 years old have the same negative attitude. And they miss original evidence constantly.

In every city I've conducted a seminar, there is always a contingent that believes "it isn't there", it's not worth looking for. And inevitably, there will be someone there in the room who says "hey, I find original evidence all the time." If ever there was a self-fulfilling prophecy, this is it.

If you want to be a truly ethical professional land surveyor, you will surgically remove the "it isn't there" attitude from yourself and those that you rely on in the field. It's that simple. Without this change, your license takes on incredible liabilities everyday.

"The evidence is there". That should be where everyone starts the job; with an attitude of "I will find something worthwhile in the field", you will greatly improve your success rate. The real keys to finding corner evidence are:

- Have the right attitude
- Be prepared by searching the record completely
- Use your imagination
- Understand the legal principles involved in corner point identification.

Good luck out there! Evidence is just waiting for the real surveyor to find and use it.

Chapter 35

SPECIALIZATION

> "Mars is essentially in the same orbit....somewhat the same distance from the Sun, which is very important. We have seen pictures where there are canals, we believe, and water. If there is water, that means there is oxygen. If there is oxygen, that means we can breathe."
>
> *Former Vice President Dan Quayle, Specialist*

Can you imagine going into a hospital for some specialized treatment and the doctor that was supposed to take care of you for your heart condition is not available? So they quickly replace him with some fellow who is a podiatrist, who specializes in working with feet. It would be a rather unnerving situation. Yes, in the medical profession they have a wide variety of disciplines. Within what's called the "medical profession" is a very complex set of specialized areas.

I think surveyors sometimes sell their profession short because they don't realize the complexity of their profession. It has become even more complex in the last 50 years. We of all people should recognize this and realize what effect that may have on our ability to serve the client effectively. There are certain basic capabilities all surveyors should have. Those would include some mathematical background and understanding of the operations of the various instruments that surveyors would use. All surveyors should have some basic knowledge of business, personnel management, and related subjects.

But let's consider the drastic changes that have taken place in our profession in this century. The surveying profession has expanded into areas that no one would have dreamed of before. We land surveyors have specialized in boundary work. We have construction surveyors who specialize in the laying out of construction plans as well as the topographic mapping of areas that are proposed for construction design purposes. We have geodetic surveyors who know nothing of those first two areas but are specialists in geodesy and high order surveying and mapping. Add to that the industrial surveyors, people who have never set foot out on the ground. Their entire career has been spent laying out machinery, checking deformation of equipment. It's a big world out there.

And now GIS (Geographic Information Systems) have come into play. Many surveyors are specialists in GIS matters now. We also have surveyors who are licensed professionals whose main direction is in computer software design and operation. In many states surveyors are also doing Photogrammetry work, a science that had never been considered prior to the airplane and the camera.

When I consider this vast array of specialties within the surveying profession I must honestly admit that I don't have a lot of training or background in many of those areas. Like it or not, specialization has taken place in our profession. We probably have some within the profession who will not admit that specialization has taken place. They may even, as a result of ego, think they already know about every one of these specialties.

I believe that would be a very rare individual. In this modern age of surveying the true professional should recognize his or her limitations. Those limitations are not an admission of failure or weakness or ignorance. Rather, they are an honest admission that one's training, experience and general interests lie in certain directions. They usually don't include every facet of the survey profession.

An ethical surveyor will know well his limitations. The true professional will seek out advice from others who know him well. This will provide insight into what they think his or her limitations are. We should weigh such input carefully.

No matter how much you need to work, no matter how tough times are, should a true professional take on work that is beyond his grasp or capability to perform? To do so is to defraud a client. The best of hopes and the wildest of dreams cannot perform a professional project. Only by luck can we succeed when operating this way.

So what are your specialties? What do you know well? And what don't you know well? And how do you know you know it well or not? How well informed are the others in your organization as to what those limitations are, where your specialties truly lie? Does everyone in your firm know that you only take certain types of projects? Or are they taking anything that comes along that has a dollar sign attached to it? We cannot afford to take that approach any longer.

This author's experience is 75% boundary work and I am constantly amazed at the poor work that I find in the field by others. I don't criticize those because they lack that knowledge. I criticize them because they went ahead and did a job beyond their true specialty. For example, these could be firms that have traditionally specialized in construction and engineering related work. That's great work and frankly there's more money in it. But for them to go out and to attempt to do a complex boundary survey is absolutely unconscionable.

I'd like to take you one step beyond. One of these days our profession, on a state by state basis, if not nationally, will have to come to grips with the specialization of surveying. It may have to be reflected in the licensing process. Just as engineers can be licensed in a couple of dozen different disciplines in some states, surveyors will also face this in the future.

I have mixed feelings about this concept. It could get out of control. I can just imagine having to pay three specialty licenses for each state I'm licensed in!

The general argument for specialty licensure finds that there should be three basic categories:

- boundary or cadastral surveyor
- construction/engineering surveyor
- mapping scientist or geodetic surveyor.

Nothing would stop someone from being licensed in all three specialties and there's really no reason for the state to charge more. It would limit what one could advertise to the public. You would have to stay within your specializations. Given the diversity and complexity of this great profession, I believe that we should have separate tests for each of the specialty areas.

Surveyors don't like change. We are a very independent group. Many surveyors bristle against the fact that they even have to be licensed at all. Let's face it, we're involved in a very complex and exciting and dynamic profession. We need to change with the times, we need to be aware of what's going on out there in the world, how we affect it and how we interact with it. It's time for the surveying profession to seriously consider specialization in the licensing process.

I offer these thoughts because I believe it will come down to that eventually. I'd rather see us design and control this process from the start. Otherwise the state governments will do it for us. Either through legislation or regulation, we could have a nightmare shoved down our throats. Why not devise our own system and controls up front? I always prefer controlling my own destiny.

Perhaps the greatest benefit to specialization licensing is this: The tests and experience requirements could be focused in on the specialty areas. It would provide us with a far more qualified group working in any specific specialty. I am ready for that, especially in the boundary area!

If you totally resist such specialty licensing, I have one suggestion. Whether we have it or not, we as ethical professionals should know our limitations and operate safely within them. That is the only way I know of that we can avoid such controls. But since I also know most surveyors will reject such "self-control", I think we better get ready for the alternative.

You can make the choice today. No matter what your state or province does; no matter how it is addressed or ignored, you can be a super-professional by assessing your real specialties, and then staying in them.

Chapter 36

EATING CROW, PROFESSIONALLY

"One of the quickest ways to meet new people is to pick up the wrong ball on a golf course."

Bits & Pieces

Have you ever known someone who was caught red-handed at making a significant mistake? And yet no matter how obvious the facts were, they could not admit they had made mistakes. Even when they are forced to admit that they are not in harmony with the way things should be, they've got 25 other excuses. They may blame others or simply turn around and attack those who have found the error.

This is a part of human nature. Every one of us suffers from it to some extent. Seeing as how surveyors are human (perhaps more human than most) we therefore are not immune from this dilemma.

I was doing a survey in central New Mexico when I came across a place where another licensed professional had supposedly subdivided a section. He was breaking out a forty acre parcel for a ranch. I, on the other hand, was breaking out another 120 acres in another area of the section. Obviously our surveys overlapped and used the same monuments at a number of locations. I noticed on his plat, (which he failed to record) that he claimed the west quarter corner was lost. So he single proportioned it in. I had to get the copy of his plat from his client because he would not share it with me.

I was rather curious about this lost quarter corner because I knew that this entire township had been brass-capped by the GLO in the 1940's. I went and got the notes for that dependent resurvey, and found that the GLO had set a brass-cap and taken two bearing trees, at this quarter corner.

I had access to some aerial photos of the area that were only three years old. I looked at them to see if there had been any activity by man in the area of that quarter corner. There was none visible. In fact it was in terrain such that there was probably very little ever to go on there. So, I sent an unlicensed technician to the field with the notes in his hand and said, "Go out there and look for that quarter corner".

He drove as close to the area as he could get and within 15 minutes had found the brass-cap and the bearing trees. They were so obvious, it was not a question of someone not being able to find them. I am convinced it was a question of someone not bothering to look. You see, all of his work was in the east half of the section and he would have had to traverse a couple miles of extra line to go get that west quarter corner. So he just made up the fact that it was "lost." He also falsified his ties to several other corners.

As fate would have it, the position of the quarter corner (as found) is significantly different from the position of his quarter corner, which he proportioned. By the way, he did not monument his portioned quarter corner. Of course not; that would have involved and required traversing out there again. So he just did it all on paper.

This was a tough situation; I had to go talk to him about it. It made it tougher because he was a member of the board of registration and a very prominent surveyor in that community. But I decided that I had to go talk to him about it. I did it as gently and as kindly as I could. And suddenly I was in a whirlwind of denial, blaming, and justification. After about 30 minutes of his tirade, I calmly said "Well, no matter why you did it and no matter what it is that you did, you have a quarter corner position that is over 100 feet different from what it should be. It has affected your entire section subdivision and your clients' property has been improperly surveyed. What are you going to do about it?"

His response was the "shock of the year". He told me that he had already completed his survey and that he had no intentions of going back out there again unless I was going to pay him to go fix it! Within a few moments I got up off the floor; I had never heard such an unprofessional and unethical attitude in my life! I asked him how in the world he thought it was my responsibility to pay for this repair. He told me that I'm the one who had the problem with how he did the survey so I ought to pay to have it fixed.

Now since he's on the board of registration (or was at the time), this put me in a very delicate situation. I truly did not know what to do. And you may not agree with what I did. At the time, I had two friends that were under scrutiny of the board, trying to get their original licenses. I did not want to jeopardize their opportunities because of some form of revenge that could be taken on me by this gentleman. So I proceeded to not complain to the board. I chose rather to inform his client of the errors found. I left it to the client to complain to the board.

Now some 8 years later, I regret not having "gone for the gusto" and nailed him. I do not know if the client ever followed through with a formal complaint to the board or just to the individual. Perhaps he fixed it. I hope he did.

The point of all of this is simply a question of eating crow. If you've got screwed-up surveys in your past, it's time to eat crow and admit it. I have several screwed-up surveys, I know where some of them are; others are unknown to me, but surely exist. I have even informed a few people of their existence and told them how to fix it. I have gone back and repaired several old surveys. I just had to learn to eat crow. I couldn't make excuses any more.

Eventually we all have to eat crow. The only question is, will we do it professionally? Eating crow professionally is an ethical action and it should be a basic part of the registered land surveyors diet.

Chapter 37

ADVERTISING

> "The right to do something does not mean that doing it is right."
> *William Safire*

There was a time in the United States where any advertising by a professional that went beyond notice of your existence was considered completely inappropriate. Surveyors, along with engineers, were aware of this "taboo" and avoided anything except a humble ad in the yellow pages. Doctors and lawyers were the same.

Perhaps you have noticed a number of the professions going all out in advertising in the past few years. Some of it has been very tasteless, especially from the legal profession. Even some surveyors and engineers have some rather blatant and boastful ads in the yellow pages, newspapers, and other journals. What changed all this?

In 1976 the United States Supreme Court made a landmark ruling on this subject. They felt the limitations on advertising as dictated by professional associations was a serious detriment to competition and an open fair market. Most trade organizations backed off from outright banning of advertising after that point. That ruling, along with some other general changes in the American "culture", is the main reason we have seen this shift in policy.

On the surface, one could assume this means there is no restraint, just a "free for all" out there. And in some professions, that's about where they've gone. Does this ruling and change in attitudes relieve the professional from any ethical responsibilities? I don't think so.

Remember, our ethical responsibilities are toward the profession and society as a whole. It is not just a matter of "what can I get away with?". You can probably get away with much; but is that where we as a profession should really be? I doubt we'll ever see surveyors running ads on TV with dialogue like this:

"Hi friends, Honest Dennis here with Acme Land Surveys, Limited. Do you need a fast survey? We have a special this month only. We will throw the plat in for free with every boundary survey done. That's on top of our already low prices. How do we do it? We work with such a volume, we save on all our materials and supplies, and pass the savings on to you! And that ain't no bull."

You know how you feel about advertisers who use the above approach on a regular basis. I go out of my way to avoid those nerds! There is nothing "professional" sounding about any car dealer or furniture hustler using these tactics. So why would an intelligent person hire a surveyor who works that way?

I have reviewed about 400 yellow page ads for surveyors across the United States. (It's one of my past times when sitting in a motel room resting up for a seminar the next day). In most cases the ads are very tasteful, and I'm sure relieved about that. But I've seen a few that really bothered me. Some of the elements of some ads I felt were tasteless or unethical follow:

- Lowest prices in XYZ County
- Phone quotes on any job
- We'll match any price!
- Best surveyors in town
- Fastest surveyors in town

How do some of those claims strike you? I'd like to see the Board of Registration investigate those folks. If they can live up to their claims, they are probably doing very poor work! Think of the impression the general public gets of the surveying community when they read such ads. Hiring a surveyor would appear to be similar to buying a used car or selecting a janitor. It does not appear at the professional level I think it should be.

So regardless of "what you can get away with", what ethical responsibilities do we have in our advertising? ACSM and some state associations have put out some general guidelines. These are worthy of your review. Many firms have brochures, or advertise in newspapers, journals, or other media. I would like to offer you the following thoughts on this subject:

- Advertising should not be boastful, bragging, or otherwise ego-inflating.
- Ads should disclose the areas of expertise and specialization, as well as location and geographic coverage.
- Fees should not be discussed or even alluded to.
- Ability to make phone quotes should result in the death penalty.

- Avoid any deception, exaggeration, and half-truths.
- Be cautious of the effect of omissions in your ads.
- Never specifically advertise (or solicit) work from a client who has already hired a surveyor to perform the same work.
- Avoid "showmanship".
- Watch use of slogans, mottoes, or other sensational language.

While one may think there are buckets of money out there awaiting the most clever and perhaps sinister advertiser, there is a bigger picture to consider. Such wild ads berate and insult our profession. They lower the expectations of the "buyer". They deflate our own self-image. I think it does far more damage than good, for anyone involved.

The bottom line with professional advertising:

"Show a little class."

Chapter 38

CONFIDENTIALITY

"Loose lips sink ships."

Secrecy slogan in World War II

We live in a society where there seems to be very little respect for confidential matters. We see the President's own staff constantly leak information to the press. There is always someone willing to disclose information which was to be hidden for good reason. And often there seem to be more rewards for those who divulge the secrets than there is punishment. We live with a populace that simply cannot keep a private matter "private". As the advertisement goes, "inquiring minds want to know!"

Most surveyors will never have to deal with clandestine or confidential matters such as the above examples, but we do occasionally deal with information that should be kept private. There are actually several aspects to this subject, and the surveyor should be sensitive to these issues. In World War II they had a saying about confidentiality: "Loose lips sink ships." Professionals should never have loose lips.

Perhaps a definition of confidentiality would be in order. Confidentiality is something requiring secrecy and privacy. It includes a level of trust and "confidence" in the hearer. The term implies the most intimate of a business relationship and it's safeguarding from outside meddling or influence. Those are powerful words!

Let's consider a few of the possibilities in this need for confidentiality on the part of the Professional Surveyor. As you read these concepts consider the ethical implications of each.

CHANGING JOBS :

As surveyors (or any other professionals) change employers, there is a great need for confidentiality. This especially addresses client information, including projects, finances, plans, design techniques, and schedules. The knowledge gained (other than technical experience and OJT) with an employer is not subject to sale, barter, or disclosure.

Unfortunately, many employers hire a particular individual to access their information regarding clients, potential projects, and internal corporate secrets. This practice is so common, many reading this may think the author is some prudish fool. But confidentiality is the foundation of trust between professionals, their employers, and their clients. Think about it: An employer who encourages you to divulge secrets of a former employer or client knows one thing for absolute certainty—you are not trustworthy. That employer will always watch you and be very wary of your loyalties. Is that a good professional ethical environment to work and live in?

COMPUTER SECURITY:

The high-tech age has brought us new ways to be dishonest and sneaky. For most businesses in America the computer has become the complete focal point for all data storage, manipulation, and operation. This wonderful tool provides all the information one could want to get on a client, a project, or personnel records.

The ethical employee would never take advantage of his access to this sort of information. Confidential information is confidential, wherever it is stored. Upon termination from a firm, an employee could steal more information on one floppy disk than a dozen other employees could do from memory. Obviously this is not acceptable behavior.

WHISTLE-BLOWING:

The term "whistle-blowing" means one is turning in another for some illegal or unethical act. There has been a significant increase in government and industry to actually formalize whistle-blowing processes. We even have "800" numbers for turning in fellow employees for improper or wasteful acts.

The advent of formalization of whistle-blowing has encouraged more people to use the process. Obviously there are some ramifications of this action. Some work places have bred paranoid employees, hard feelings, and lawsuits to recover from whistle-blowing accusations. And the process has been abused, mainly by those who use the system to harm others with fabricated accusations.

There are some serious considerations for the employee contemplating blowing the whistle on his fellow employees or even his employer. Confidentiality comes into play and cannot be cast aside. If you are going to take some kind of action, you have the right to demand confidentiality of your report. However, when the offending party denies the accusation, your confidentiality may disappear. Therefore it is very important you weigh all the ramifications of your action.

It is also important you be sure you have gone through the proper channels prior to taking a whistle-blowing action. These are very difficult situations. Is your action going to result in positive performance, or just a muddied relationship at work?

Even more important are the confidential issues for the employer. When an employee comes to you with a real or perceived problem, how will you handle it? Will you treat it with the privacy it deserves? Or dismiss it completely? Or perhaps throw that employee to the wolves? This is the sort of issue that makes a manager sweat blood. Whether the complaint or concern is done through a formalized process or just over break time, what ethical responsibilities do you have to that employee? To the other employees?

Employers have a responsibility of confidentiality to their employees. It could be compared to the confidentiality between a doctor or lawyer and their clients. It is that serious! Confidence in you as a person, an employer, a supervisor, usually rests with how you deal with difficult and sensitive information. It will test your mettle. Ethics will get you if you don't respect them.

The ship you sink will usually be your own. Tighten you lips and action.

Chapter 39

"DOING IT YOURSELF"

> "An attorney who represents himself has a fool for a client."
>
> *Some lawyer who lost his own case*

We live in an age where many of us build it ourselves. We fix it ourselves. We pump the gas ourselves. It has become a do it yourself world. There are several reasons why this has occurred in our society. One is that we enjoy doing these things....we need challenges or opportunities to succeed at something outside of work. So we build decks, add rooms, repair faucets, and change our own oil.

Another valid reason for this change is simply the cost savings. It is usually cheaper to do things for ourselves than to hire someone to do it. In the United States and Canada, we have almost been engulfed with a passion to save a dollar; so you do it yourself.

For most activities this is an innocent and profitable practice. Saving money and using your own skills and talents can be rewarding in many ways. But for the Professional Surveyor, there is one area we need to consider.

Should you survey your own property? I know many who have done so. Perhaps it was just an "improvement location certificate" (or your equivalent) for a mortgage or refinance. Perhaps it was a parcel of land you were planning to develop, and a "free" survey would certainly add to your profit margin when you sold the lots.

These desires are certainly innocent enough. But are they ethical? The conflict of interest raises it's ugly head once again. And at a minimum, the "appearance" of a conflict of interest is at your side. Should we be doing our own surveys?

There is a common saying among lawyers that goes like this: "An attorney who represents himself has a fool for a client". There is some wisdom in that saying that we could glean. If a boundary dispute were to erupt between you and a neighbor, your actions and approach could be called into question. How would this "appear" to the average citizen...say a member of a jury perhaps?

At first such an accusation offends us. If you are like me, I would do the survey in the same manner regardless of who owned the land, even if it was me! But that is not the issue here. The issue is whether you were being very smart to do your own survey at all. "Appearance" of a conflict of interest is a long way away from a real conflict of interest. This problem could be called into question for a survey you have done for a relative, business associate, or friend.

Let's face it, in our suspicious and often accusatory society, there will always be an "appearance" of a conflict in somebody's eye. You cannot get away from it. But many of these appearances are easily swept away by common sense and reasoning. But some are not so easily dismissed. And it is the latter that could cost you a fortune in court just to "prove" there was no real conflict to begin with. Is it worth it? You must be the judge.

Generally, where a professional has the "opportunity" to establish a boundary, volume, or the location of a structure that would be to his financial advantage, there is the appearance of a conflict of interest. Some examples follow:

• A vague call in a legal description allows two possibilities in interpretation. You may have always chosen a certain answer, but the fact that you did it on your own property raises serious doubts.

• Your best friend is now a client doing some major earthwork. You are doing the quantity estimates and you must estimate volumes. Perhaps you always "round up" or estimate in a certain manner, but to do so in this sort of case leads some to speculate you have shown favor toward your friend.

• Your firm has prepared the construction plans for roads, buildings, and parking lots. You begin to stake these and discover a fundamental error in the plans. You change the staking, location, and arrangement of facilities to fix (and perhaps hide) this error. Perhaps you end up causing the structures to violate zoning setbacks, etc. While you were trying your best to keep the dozers and "yukes" moving, you have inadvertently strayed into a conflict of interest that may look worse to a judge or jury than it really is.

It does not seem fair, does it? But in this litigious and almost paranoid society, nothing is fair anymore. The wise professional will see these potential problems ahead of time, and will work to avoid them.

Sometimes it is so innocent, you don't see it coming. I did the lot survey for my neighbor. They were moving into the neighborhood having a new house built on the lot next to mine. As a gesture of welcoming them, I offered to do the location survey for free.

Besides, I reasoned, I already know where half their corners are already...they were my corners! When I did the survey, one of our common corners was lost. I proportioned it in properly, and it indicated the fence was encroaching onto me. I told him we would need to fix that some day if and when we ever rebuilt the fence. We get along fine. But the day he sells that place will be the day I find out how far I really went in this ethical menagerie.

Not until years later did I think about the potential embarrassment, or cost, of such an action. I hate not being able to be a good neighbor, or just being a "nice guy". But the Professional Surveyor must recognize that his or her stature demands a higher awareness and level of caution. This is for his own good, and that of the profession.

How can we avoid these situations? First, we need to remember to Think! Then perhaps we can find other solutions for these "appearances" which will assist us in not having to turn down too much work. For instance:

- If we are doing work on our own land, or that of a friend, associate, or relative, perhaps the concept of "peer review" could help us. Could you have another professional review your work? This could be documented and made a part of the file, if not the plat itself. Be sure to select a properly neutral party and pay this reviewer. To fail in these actions could result in more problem than you are trying to avoid.

- If you are working on a project which has potential conflicts, ask another professional in your office or firm (if possible) to actually take the work on. Let him or her be the "primary" party in charge of the work. This could avoid a conflict and may even raise your ethical stature of your peers at the same time.

- Be honest and open. If you find yourself lying, it isn't worth it. The professional needs to assess the benefits and risks, and then make the appropriate decision.

COMPETITION

> "The nice thing about teamwork is that you always have others on your side."
>
> *Margaret Carty*

This is America! Competition is like a birthright in this country. Just look at our captivation with sports events, game shows, and movies with that theme. Competition is what makes the United States run. It keeps the world turning—or so we think.

There have been debates for centuries as to the "good, bad, and ugly" of competition. In particular, sports have seen some unfortunate events that have marred the activity. Occasionally a player or coach conducts themselves in a very unsportsmanlike manner. Their action brings shame and dishonor to the sport. But they usually did it in the name of competition—they want to win at any price.

We sometimes find this same sort of thinking in our professional and business lives. We all have competitors, and we want to win. We want all the work, all the profits, all the accolades. Our egos often drive us far beyond where our brains would have. Competition is a way of life in the western cultures.

When competition is balanced and ethical, it is a very healthy thing in the business world. Our society benefits greatly from the side effects of competition. Consider where there are monopolies in place that prevent any new thinking, price reduction, or product expansion. Some classic examples are government (especially the Post Office), AT&T prior to deregulation, and many utility companies. They have little motivation to improve service, expand markets, reduce overhead, or redesign their operations. Competition has some good to it.

But competition does get out of control. Let's focus on you as a professional surveyor, and your relationship with other surveyors in your community. How can competition get out of control in these relationships?

One of the more common problems with professional competition is the inclination to "bad-mouth" your competition. This is done in a couple of ways.

First, open mockery of the competition in any setting is done by some. They reason that if they bad-mouth them long and often enough, they will lose their business and you will get it all! Secondly, there is a more subtle (and more common) approach. It is done with new or potential clients and involves insinuation, suggestion, innuendo, and clever story telling which undermines your competitor.

Neither of these approaches are ethical or professional. It may get you more work in some cases, but it sabotages the profession as a whole. Do we really want surveyors to be viewed as you would view competing waterbed stores, phone companies, or lawyers who chase ambulances?

But does this practice of bad-mouthing really get you more work? Do we really think the American public is that dense? I for one abhor a business that purposely ridicules its competition. I find it insulting and offensive. The only clients you will "impress" with this tactic are those who treat people similarly, and are your clients only as long as they think they are saving a buck. That's not much of an endorsement.

Another form of bad competition is through the manipulation of pricing. This problem is discussed in more detail in the chapter on "Low Bids" and "Price Fixing". Suffice to say that trying to ruin or eliminate your competitor through lowering your prices is one of the most ludicrous and self destructive actions any professional can take. It is desperate, it is foolish, and it is wrong. If your business cannot survive because of your competency, professionalism, and range of service, then maybe you should not be in business! It's that simple. The American way does not allow for such stupidity, and our profession should not be plagued by this absurd disease.

Treat your competition with respect. Whether they return the favor is not the issue. If you operate professionally, you will be able to hold your head high. If you also operate ethically, you will be treating them and yourself with respect. That is what counts.

Chapter 41

RECOMMENDATIONS TO THE BOARD

"Maturity is the balance between courage and consideration."
Stephen R. Covey

Once you get your license most surveyors get into a situation where they start getting letters of recommendation to be sent to the State Board of Registration. Some person that they know or that they've worked with or had worked under them needs a reference. There's nothing really wrong with this process but I'd like to throw out a couple of thoughts regarding this.

Many people take this as an opportunity to return a favor to someone. They send in a glowing, beautiful recommendation to the Board. I've personally known many people on the Board and most of them take this process quite seriously. They would like to see your candid, honest responses and answers to the questions that the form letter usually asks. Yet the vast majority of the time they get glowing reports.

Now obviously, if you are a potential registrant, you're not going to ask for these recommendations from people that you know would give you a poor recommendation. But I've personally been faced many times with that ethical dilemma. Here's somebody who I work with now or who's my friend or who's been a fellow worker in a surveying association or something for years and years and now he needs me to do him this favor. I must admit that there have been times when I have withheld information that may have kept that person from being registered. After a few years I really started thinking about it and I offer these thoughts from an ethical perspective. Do we not owe it to our profession to be honest and sincere? Yes it's hard, it's tough to write the truth if the truth is somewhat negative about a person. Especially someone that you do appreciate and do work with and have known. Especially someone who's actually given you the honor of giving a recommendation of them. But the Boards expect honesty and openness and we should expect that of one another in this profession.

I am sure that 99% of the time a negative response to one of those things will not stop someone from getting licensed. But it may cause the Board to ask some questions, to ask the right questions. They may focus on certain aspects of professionalism or technical knowledge when interviewing or testing that applicant. It's all for our own good, for the profession as a whole, that we be honest about our recommendations.

So next time you get one of those forms, it could be real easy to go through and check off outstanding for every category that they ask. But try to put some thought into it. I would recommend that you let it sit for two or three days. You need to think about it; thoroughly review in your mind who that person was, what their good points were as well as their bad points. Then try to condense that on to the form in such a way that it's an honest evaluation of that person. I have to admit that a couple of times I've let some real losers slip through. I'm sorry about it now.

There is a reason for this process. Let's do our part when requested. Let's be honest and open, for everyone's sake, especially for the potential registrant. This is a major ethical issue!

Chapter 42

RESPONSIBLE CHARGE

"Where were you when the you know what hit the fan?"
Common question asked by Management

Most states have language within their registration laws that require registrants to be in "responsible charge". There are two general divisions of this concept. Registrants are to be in responsible charge of the work they sign. But also, applicants for survey licenses are often required to submit proof that they were in "responsible charge" of work they performed. Let's look a little closer at these two areas.

In most cases the individual states have defined the term "responsible charge". These definitions have some common threads in them across jurisdictional lines. Generally the elements included are:

- Personal responsibility of the work
- Complete control and direction of the work
- Knowledge of all done under their direction

There are some discrepancies in the use of the term responsible charge. If a registrant is supposed to be in complete charge of the work he signs, how is it that an applicant is also required to be in responsible charge of this same work? Some of the state's definitions need to be modified to correct this ambiguity. This problem will not be dealt with further in this text, and we shall limit the discussion to the registrant as he or she practices.

The concept of responsible charge does have some significant ethical implications. In particular, the Professional Surveyor who is signing plats, maps, land descriptions, and other reports must understand those implications. Your signature and/or seal on the document is prima facie evidence that you claim to have been in complete charge. You certify that you made all the professional decisions; you were aware of all the alternatives; you personally knew the steps that project (and ultimate product) went through. Ask yourself, is that really true?

Even in a "one-crew shop", it's hard to make that statement. And many of you know how difficult it is when you are running a multi-crew survey organization. Things can get out of control rather easily.

This author has had "opportunity" to run four crews at once under his license. I know many of you daily run far more than this! I must tell you what a nightmare that experience was. When those crews were running all over the county simultaneously performing critical work, it was far more than I could do to say I was truly in responsible charge. They were staking large construction projects, subdividing rural sections, doing lot surveys, performing control work for photogrammetric projects, and performing topographic survey work on future projects....all in the same week. And on top of that, I had office staff drawing plats, writing descriptions, and composing reports. All this under my puny license!

What are the really significant points in a project where the registrant really needs to make the decisions? As stated earlier, it is whenever a "professional" decision is made. One must read the definitions of "professional" in their state regulations. While they may vary between jurisdictions, the process for determining professional from non-professional decisions can be sized-up like this:

The whole purpose for licenses is to protect the "safety and welfare" of the public. Your state license laws probably have words to that effect. Therefore, any decision that even remotely comes in contact with this concept must be made by the professional. And remember: A decision made without knowledge of the alternatives is not a valid decision.

Some projects are very technical in nature. They do not require a large amount of professional decisions, opinions, or design. Rather, they are functional. I would include most topographic projects, photogrammetric jobs, and even most construction staking operations. Even with these functional activities, the registrant is overloaded unless he or she has the right crew chief in charge. Having the right people in the field is of paramount importance to any size organization. But what about the other types of projects?

Having done some geodetic control work, I know that much of it is technical in nature. But some professional decisions can crop up when performing such tasks. Large-scale GPS jobs often require decisions on how to use the existing network. The NGRS (National Geodetic Reference System) simply cannot "cope" with the type of data we now use from GPS. Further, mapping control as it ties to cadastral control requires professional judgment.

But in my mind, the most significant area where the registrant's "responsible charge" is most often compromised is in the boundary survey arena. Even with one crew, the registrant stuck in the office is hard-pressed to really know what was going on out there. I've been in the situation where my crew chief did not communicate properly with me. He thought it was below him to have to run everything by me. I really did not want "everything" run by me, but corner evidence issues and contacts with adjoiners were of importance to me. I ultimately fired this person...I simply could no longer trust him.

I have had the privilege of having some really excellent crew chiefs work for me too. And having a registered professional surveyor for a crew chief is a luxury I am currently enjoying. But no matter how good your help is (and my current help is the best I've ever had), and whether they are licensed or not, does not relieve me of my responsibility to be in charge of the work.

I've known many surveyors who run multiple crew operations for large civil/survey firms. They never get to the field, never even get the chance to ask all the right questions of that crew. They have to rely completely on that crew chief. Again, trust in your people is a wonderful attribute. But it does nothing for you and your liability when you have a problem. The prudent surveyor must have a review process of what is going on. Sheer volume of the decisions being made eventually will catch up to you.

I have watched surveyors, engineers, and other scientific professionals get roasted on the witness stand for not knowing what was going on with their subordinates. The worst excuse you can use on the stand is "I was not aware of my staffs actions". Only a few weeks ago I saw an attorney make a big deal about a cost estimate made by a technician. They had the engineer "in charge" on the stand. His assistant had released an estimate of $5,000 for a repair to some damage. The engineer raised that estimate to $36,000, a rather significant difference! The attorney attacked the engineer's lack of being in charge of this person. This testimony helped to lose the case.

Ask yourself the question: Are you really in responsible charge of the work under your direction? Loyalty and trust in your employees is noble, but irrelevant in a legal action. It is irrelevant (or perhaps even damning) in an investigation by the Board of Registration.

We as a profession need to search for ways to enhance our responsible charge capabilities. I would like to offer you a few ideas to get you thinking about this:

- Bigger is not always better. Some survey firms think they are a much greater success with ten crews than two. Is that really true? Most are not making more money; just passing more money through their checking account. We give up control and "sense" when we get too big. Is it worth it?

- Hiring the right people is really a big boost. Good staff does not relieve the Registrant from his charge. But it should relieve him of non-professional duties that eat up his time. The good manager will have people

122

doing the non-survey tasks with minimal input from him or her. Too many "managers" think they must have control over everything. They waste valuable (and chargeable) time on payroll, supplies, or composing letters. We should hire competent people to do those things! If you "work smarter", you'll increase your time for the assignments that require your professional judgment and control.[1]

• Institute peer review. There is a chapter on that subject in this book. Discover ways to have another set of professional eyes look at your work— and return the favor.

• Consider splitting the responsible charge with another registrant. There are some limitations to this from state to state, but worth exploring. Another person in responsible charge will greatly reduce the taxing of your time. This option does cost more money, especially for the small firm. One must weigh it in light of the risks your firm is taking.

Your ethical responsibility goes beyond what the state may specifically describe in regulations. I encourage you to consider how much control you really need to have. Consider your liability, your responsibility to the client and society at large, and your own personal conscience. It would not be good for every registrant to ask him or herself at the end of every work week:

Was I in responsible charge of all I did this week?

---

[1] Another element of working smarter is to fire those employees who consistently add to your workload. This can be as much a drain on your ethical and legal responsibilities as any other issue.

Chapter 43

ENVIRONMENTAL ETHICS

"Leave no trace of your passing."

Ancient Ute Indian principle

Surveyors are in the business of helping to develop land. We map it, we stake it, we parcel it, we "survey" it. And with land development, there are inevitably some who feel there has been damage to "Mother Earth". I do not consider myself an "environmental whacko". But I do care for the land and resources tied to it.

Surveyors cannot stop the tide of development. We are simply a part of the process to make it happen. You may want to avoid work that violates your own personal environmental ethics, but few surveyors ever turn down work! We are a part of the process, that is why we exist.

But in a more subtle way, there are some important and significant steps we as a profession can take to display our environmental ethics. Let's look at this a little more closely.

Do your crews "tread lightly"? Obviously this would depend on the type of project. If the activity is going to result in the clearing and scarifying of the land, I guess it does not matter. But there are many high-quality developments being made across this country.

Surveyors are well known for their lack of environmental ethics. They have chain sawed trees that needed to stay. They have driven the truck through wetlands, meadows, and dry lakes, only to leave a hundred year legacy to their imprudent actions. Surveyors have drained the oil in the company rig in a field on a job, and have painted fluorescent colors all over walls ("TBM") and sidewalks. Can't we as a profession take a little more pride in our work and the side effects of our presence on a job site?

I've seen construction staking do almost as much damage as the road itself produced. Flagging, paint, stakes, nails, rebars...all tools of the trade; but could we be a little more sensitive to those who live right there on site at the time? I watched a topographic project in the forests of Arizona. The surveyors were supposed to tie in and show every existing tree so the design could go around them as much as possible. Yet the surveyors actually chain sawed 30% of the trees just to perform the survey. This could have been avoided, but the boss did not want time wasted in additional "set ups" of the instrument. I cannot imagine the cutting of over 25 trees to have saved time, but the end result was a sorry excuse for a professional job.

I have run behind some survey crews that left more "evidence" of their survey by the trash they left behind than by survey evidence. When working in the field, show respect for the land, client, and the adjoiner. More importantly, show some respect for yourself and your profession.

This short chapter is here for one purpose: to have you ask yourself if your surveying firm is showing a sensitivity to the environment. There are environmental ethics for the professional surveyor to consider. As the sayings go for public lands, "tread Lightly", and "Leave no trace of your passing." Please think about it. Surveying does not have to be synonymous with destruction.

Chapter 44

PEER REVIEW

"In the multitude of counselors there is safety."
Proverbs 11:14

Over the twenty-four years of my experience in the surveying profession I continue to be amazed at how complex and legalistic our tasks have become. This is not to say that they shouldn't have always been properly done but it seems that there is a greater awareness to get the job done right. That may be in part because of potential liability and litigation.

But it seems the majority of projects that I have worked on (both large and small) in the last ten years have been very complicated. Always filled with complex legal research and corner evidence analysis. I continue to see more and more complicated situations regarding easements and the possible reversions of those easements. The effect of unwritten rights and prescriptive easements has never been so great. It is a complex world out there and if the surveyor wants to do the job right it's going to require a continued increase in complexity and attention.

Because I speak so often in my seminars about the Public Land Survey System (PLSS), I will use it as an example. When I consider the complexities of that system I am amazed at the amount of effort it takes to get a job done. Consider the fact that just since I've started surveying, the PLSS is twenty-four years older. I often wonder how much evidence has disappeared in those twenty-four years? How much more will vanish in the next fifteen or twenty years of my career? Every day that goes by the evidence gets older. The records get more complex. More government regulations and complexities are added to a potential project area. The surveyor draws further into more difficult and potentially liable tasks with each passing year.

It would seem therefore, that the ethical surveyor would be wise to introduce the concept of peer review. Peer review is simply the requesting of another registrant to review your work. There's a tremendous amount of wisdom and safety in this process. Another set of eyes looking over my plat and my procedure has caught a number of silly little mistakes. It has even caused me to go back and look at the process I used in the field. I have deeply appreciated those fellow professionals who have provided me with peer review.

And on the other hand, I have provided peer review to a number of other land surveyors. I have tried to be fair and reasonable. A number of times I have spotted something more than just a "typo." I have sometimes pointed out an improper procedure that had been used. Perhaps they had tied to the wrong corner. It's good to raise questions that cause a surveyor to go back and look closely at what he did. Perhaps he would validate what he had done after my input, but the input he got from me was of great value. They've all told me so.

The professional surveyor can set up a peer review system rather easily. If you have more than one registrant in the office of your organization, you can certainly use that person. Some of the best technical conversations that I have ever had were with fellow professionals in my office. Sometimes it seems we're a little harsh on one another but I look back at that with a very positive feeling. Peer review is a "safety net" that I cannot afford to live without. Surveying gets up on the "high wire" too often.

If you don't have other professional surveyors in your office then it's time to seek out professionals whose judgment you trust. These could be people from your own community or perhaps from another adjoining geographical area. Either way it must be someone that you trust and you feel is an expert in the subject area. You can easily set up an informal agreement between you and that person. You can trade peer review opportunities with each other without a charge.

This is a subject where our professional paranoia can come to life. Often we have great fears of having other professionals look over our work. There's a fear that they might discover that you are not perfect. Well I have good news for you; they already know you're not perfect. I'm not perfect and that is the very reason why I desperately need peer review.

It should be kept in mind that there are some specialty areas in the surveying profession. I have found myself going to certain specific people for certain types of projects because I trust their judgment. Perhaps I've had experience retracing their work and I know that they know what they're doing. For instance, if I were to deal with a riparian issue in the PLSS I would seek help! I have two or three names that I would contact in various parts of the country to question them and get advice as well as have them take a look at the procedure that I am using before I finish the project. There is certainly nothing that would keep you from being able to have multiple peer reviewers based on their own areas of expertise.

The song says, "No Man is an Island", and it certainly is true. Surveyors can sometimes find themselves operating as islands either by choice or by default. But the wise land surveyor will swim to the "ethical shore". He will open up his eyes to the rest of the world and to the realities of this legalistic society. He will strive to find the right people and the right situation to set up a peer review arrangement.

It's not only for your benefit but for the benefit of the public in general. It is especially useful to your clients. This is a true question of ethics and I encourage you to find a way to set your own system of peer review.

Chapter 45

RUBBER STAMPING

> "Responsible charge means direct control and personal supervision of surveying work."
>
> *New Mexico Statutes (61-23-3 F)*

The practice of stamping someone's work other than your own with your registration stamp is called "rubber stamping". It implies that you have no knowledge of the work, or have made a cursory review, but were never in responsible charge of the work. You will find a chapter in this book on the subject of responsible charge. Rubber Stamping could fall under that subject, but I wanted to briefly cover it separately.

I have known of some rubber stamping operations over the years. It has surprised me how often a registrant is willing to sacrifice his/her entire living by engaging in this illegal practice. Some of the reasons for doing it have shocked me. Here are three I've known of personally:

• "I stamped for this kid because his dad and I were partners and good friends". Now there's a great reason to violate the law, cheat the public, and endanger your livelihood.

• "The guy couldn't pass the LS test. It's such an unfair test, I wanted the guy to make a living." Wouldn't common sense tell you this is the last person on Earth you'd want to rubber stamp for?

• "That area needs a surveyor in the local community, so I offered to stamp for him until he gets his license." This particular case involved the "stampee" failing the test 4 times, yet advertising his name in the yellow pages and local newspapers as a Land Surveyor and Civil Engineer. He was neither, but the "stamper" was both.

On the latter case, I personally filed a complaint with the State Board. They claimed they were powerless to do anything with the "stampee". But the "stamper" was contacted and investigated. He claimed he reviewed the work completely and was truly "in responsible charge". I knew this to be a bald-faced lie. The ads never mentioned the stampers name. They fully implied the stampee was licensed in both disciplines. The Board dropped the case, satisfied that this unethical liar was just being picked on.

I proceeded to review some of the survey work, and found it to be totally illegal and incompetent. It was what we call "1320 Specials". I went ahead and informed their clients of the work's unacceptable nature. I did all I could to interfere with his business. I informed local governments and realtors that this was an illegal and incompetent operation. There was no defense they could make against me. I admit, I was "bad-mouthing" them, but I did not care. I did not see these guys as competition; they were scabs on my profession.

One of the surprising events connected with the above story was the defending of both parties by other registrants. I received scathing letters from others for my actions trying to force the stampee from business. It seemed like a "no-win" situation. When the Board won't take action, I don't know what else a guy is supposed to do!

Rubber Stamping is illegal! The only defense is to lie about it. This practice is just plain stupid. If you are stamping for others in this fashion, you have degraded and disgraced your profession. You show you operate with little or no ethics, regardless of your reasoning for doing it.

If you are the recipient of rubber stamping actions, you should never get a professional license. I will work the rest of my career to deter such a person from ever joining my honorable profession. It blows my mind I have to write a chapter on such an unethical issue, but it really is out there. Let's "stamp it out"!

Chapter 46

CHARITY

"There really are only two ways of life; the way of give, and the way of get."

Herbert W. Armstrong

Surveyors are not wealthy, for the most part. There may be some "trustfunders" out there or some who gained their wealth through some means other than surveying. But that does not count. Surveyors are generally in the middle income brackets of this western world.

Even the poorest person in the United States is wealthy compared to those in other countries. It is a matter of perspective, once again. It always helps to step back and see the big picture. And this picture is one of surveyors sitting in the mainstream of the wealthiest nations on Earth.

Elsewhere in this book the subject of corporate responsibility is addressed. That subject includes the awareness of the community's needs as well as those of one's own employees. There is a responsibility for successful corporations of all types to give back to their community, even if it has no direct benefit to their business.

This brief chapter will address the need of the individual professional to give back to his or her own profession in particular.

The surveying profession is in a significant state of change. Across the U.S. (and many other nations) we see stricter licensing regulations. Mandatory continuing education for the licensee as well as 4 year degree requirements for the hopeful registrants are becoming common. Technology continues to challenge us with a fast-paced march toward smaller, faster, more precise, and more complicated gadgets. It is easy to get caught up in this "spiral" of activity and demand and forget the needs of some of those in the profession. In particular, what charitable actions can we take for our profession? Can we give something back?

The cost of college education is skyrocketing. There are few antique surveying instruments available for the public to see and appreciate in museums. Many non-profit organizations are in need of surveys, design, or other efforts which we provide. We of all people should be actively involved in the promotion of our own profession. We should care about it's historical preservation, as well as the plight of those who will carry on this great profession after ourselves. How important is this profession to you? Is it valuable enough that you could share a little time, money, or thought to it's maintenance and promotion?

Recently the NSPS Foundation has been established. Here is an organization designed to promote and sustain this profession in all of these ways. Could you afford to commit to an annual donation to help this noble effort get off the ground? Here in Colorado we have a statewide effort to get the Metro State College (in Denver) surveying program back on track again. Our professional association stepped out in "faith" and committed to a $60,000 donation. This bold effort needs every professional in the state to take up the cause and push to make it happen. Many other states have similar efforts for educational programs.

I especially am motivated by funds that are used for grants or scholarships for young people getting into land surveying. At this writing I have two of my children in college. I personally know the cost and pain of trying to fund this effort. It helps me understand what the young surveyors are facing. Excellence in educational efforts should be rewarded. We should also find ways to reward those in financial need who have proven their attitude and aptitude for our profession.

It is said that it's "more blessed to give than to receive". Whether you are wealthy or not, you owe something to this profession. No, you may not make a lot of money, but surveying *has been good* to you, hasn't it? It's time to pay up....for the rest of your life. Learn the way of give, in this and many other aspects of life. It really is rewarding, for you and those who receive.

Chapter 47

THE FEES YOU CHARGE

"If virtue and knowledge are diffused among the people, they will never be enslaved. This will be their great security."

Samuel Adams

In any profession, fees can be the source of many an argument, lawsuit, or lost clientele. Surveyors are just like anyone else. We are trying to make a living, and maybe even make a profit. But we can also see opportunities to "put it to" someone. Many of you could tell your own stories on this subject, I'm sure.

There can be the tendency to charge more from a client who would have no trouble paying the higher fee. I have seen firms charge extra hours to accounts, pad other expenses, and add in unnecessary items simply because their client was a wealthy individual or a large corporation. I've heard the excuse, "they can afford it", or "they'll never miss it".

I've seen government projects get billed for things never done. Or the old trick of charging for a certain level of expertise, but sending a technician to do it. I experienced a case where I was the project surveyor on a street and utility construction project in a remote southwestern city. The civil/survey firm I worked for needed to have a materials inspector for the asphalt, concrete, and soils. I had some minor experience in these areas, so they had me do it in addition to my surveyor job. The city was being billed for having a real inspector, but I was a bungling amateur.

Further, the company billed the city twice for my time, once as surveyor, and once as inspector. Since the job was 200 miles from my office, the city also paid twice for my travel and per diem. It was not an ethical arrangement, to say the least.

Have you ever worked on a "cost plus" job? This is where the client is paying time and materials on a unit basis. I saw the City of Phoenix get ripped off royally during some of their floods along the Salt River. We were told to "take our time" on those jobs, and we did what we were told.

The preceding story is an example of taking advantage of a client. Either you take advantage of their trust in you, of their urgent situation, or their ignorance of reality. But no matter what, it is dishonest. We made lots of money—even had a good year for the profit sharing fund, but it was dirty.

Perhaps one of the more greedy examples I saw of taking advantage of someone's urgency was where a landowner had sold his home on 3 acres. It took him several months to finally sell it, and when he did, he went through all the "hoops" to get the deal closed. The buyer had a very strict time line which had to be met. On the day of closing, the title company panicked and demanded a "mortgage survey".

I am not a big fan of the mortgage survey. They go by different names in each state, perhaps "Improvement Location Certificate", or "Property Inspection". But in most states, they are just a rip-off of the client. Society thinks they are surveys, but they are nothing more than "drive by shootings" in most cases. Such was the case in this example.

The title company said there would be no closing without a "survey". The landowner called a survey firm in town to do the survey. When the surveyor realized the urgency, he quoted them a price: 2 points. This surveyor had surveyed the parcel and house only two years earlier. He sent a single person out who literally drove by the parcel, noted there were no changes on the property, and produced an exact replica of the previous "map". But he charged him 2 points. The price of the house was $150,000, and that was in 1986. A $3,000 fee from a surveyor who did next to nothing; that is not ethical.

I personally knew both the surveyor and the landowner. I had sent over $50,000 of work to that surveyor over the previous 3 years because I was not able to perform it. After this incident took place, I never sent him any more work, and I never gave him the time of day either.

One final area to be aware of is that of the "contingency fee". These are fees that are directly tied to the outcome of the work. For instance, a surveyor does some work on the contingency that he finds errors in other surveys that are beneficial to his client. If he finds errors, he gets paid. If he finds none, he gets nothing. This sort of thing has happened.

Another example is that of the surveyor who asked for 10% of the value of land he found "available" to his client through adverse possession claims and quiet title suits. He was so desperate to find things he literally made some up. They even won a couple of quiet title claims because they picked on some elderly uneducated people in a remote area of New Mexico.

One's professional judgment could certainly be clouded when working under this sort of arrangement. I find it too much the appearance of a conflict of interest, and often a completely unethical arrangement. My advice: avoid shady deals involving contingencies.

In most states, the Board of Registration (or equivalent) is bombarded with complaints against surveyors. And did you know the majority of the complaints are about the fee? I understand that many times the client is so uninformed about the survey process, that they panic at any price over a couple hundred dollars. (The infamous mortgage surveys have helped set that paradigm). But in many cases, the people did not get what they paid for. They were often completely ripped-off.

Watch your fees. Be fair, consistent, and honest. You cannot lose with that as your policy and motto.

Chapter 48

BELIEVE IT OR NOT

"If this is coffee, please bring me some tea; but if this is tea, please bring me some coffee."

Abraham Lincoln

Many of you will remember some thirty years ago the "Ripley's' Believe it or Not" trading cards. Each of these cards contained some amazing historical or scientific facts that would stagger the imagination. Supposedly they were all true. But some of them were very difficult for the reader to truly believe and so they simply challenged us to "believe it, or not." In the mid-seventies the Ripley's' people created a television program which lasted two or three years on the same type of subjects.

I would like to create my own version of the "Believe it or Not" program. This chapter is a short collection of short stories about surveyors I have known or have known of. I have taken some effort to validate the stories in the situations that I was not directly involved. I feel comfortable that the stories that I'm about to share with you are all true.

A friend of mine who was a surveyor in Southern New Mexico was retracing a long time local surveyors work. He continued to have rather significant distance discrepancies between his survey and that of the previous surveyor. There was a consistency to it; the resurvey was always shorter in distance than the previous survey.

After two or three days on this project my friend finally realized what the problem must have been. Everything on the plat of the first surveyor was a slope distance. Nothing had been corrected to the horizontal. My friend went out to verify this fact between several points and it was obvious that this was true.

Puzzled by this strange means of reporting one's distances my friend made contact with this surveyor. He was quite concerned since the other surveyor had been surveying in the local area for some forty years. He had a brief discussion with this land surveyor over the phone. The surveyor explained to him the reasoning and the process and it went something like this: "You don't have to reduce distances from the slope if the line goes up over the hill and comes back down the other side of the hill because these two slopes will off-set each other".

Anyone who has a basic understanding of surveying and trigonometry knows that this is the most ludicrous statement ever made! And yet here is a long time, licensed surveyor who did not know the basics of reducing a slope distance. Somehow his understanding of trigonometry is so narrow that he honestly believes that the going up and back down the hill somehow compensates for slope. In fact it doubles the error. The question one wants to ask here is, how did this guy get licensed and why is he still licensed?

The next story begins in a small town in Arizona.

I had been in the court house several weeks earlier. I had made a copy of a recorded mylar plat. It was of a section subdivision for a parcel in which I was also working. I used this platted information to see what another surveyor had done in the area. I noticed several errors on his plat, some of which seemed to be typographic in nature. There were also a couple errors which were process errors. Needless to say my survey did not agree with his in every way.

About two weeks later I went into the same court house to validate another piece of information on a second sheet of that plat. I happened to glance at the first page and was shocked to see a number of changes on the plat that had not been there only two weeks previous. Someone had come in with a rapidograph pen and made changes to the plat, correcting some of the very mistakes that I had found on that plat.

There's only one person that I really think would have done that sort of thing and that would be the registrant. So I called him and asked him about it. He openly admitted that he had made those changes. I asked him how he could justify going into the court house and changing a recorded document. And his answer was quite simple. "That plat belongs to me and if I find an error on my plat I have the right to fix it from now until the day I die!"

Apparently he had never heard of an amended plat or a correction plat. Needless to say, I get very nervous with any kind of a document recorded in the court house when the very profession who has recorded that plat has representatives who feel it's their right or duty to go in and change the record.

The third story begins in 1982.

I was asked to survey a portion of a homestead entry survey (HES) near the south rim of the Grand Canyon. An HES survey is a metes and bounds survey within a National Forest which is segregated from the public lands survey system in the area.

This particular HES is the only piece of private land at the south rim of the Grand Canyon. If you've ever been to that National Park you may re-call a small valley wherein is located all of the fast food joints, most of the motels and a number of other tourist-trap related businesses. All of those parcels sit inside this one hundred and fifty eight acre parcel of private land.

The project I had been assigned was to investigate a possible encroach-ment of a new Best Western motel. The local ranger for the Forest Service felt that perhaps the new wing was encroaching on National Forest Land and he wanted to have a survey to verify this.

I proceeded to do the survey but quickly found a very strange set of cir-cumstances. At every HES corner was a well set stone with proper marks and every corner had bearing trees (witness trees) referencing the corner point. As my custom is I located the stone and then checked the ties to the bearing trees. The ties were significantly different distances than what the records said.

If you've ever worked with HES's you know that they are perhaps the most well monumented, well documented, and measured surveys in the whole system. So I was suspicious that something was wrong. The bear-ing trees on all but one of the twelve corners of that HES did not fit the calls in the record. So the question was, are the bearing trees incorrect — or had the stones been moved?

The stones appeared to be the original monuments and they were prop-erly marked and facing the correct direction. But in further investigation of the field notes I discovered that there were memorial stones set below the monument stone. These were small rocks usually the size of your hand or smaller with a cross on them. These were supposed to be about six inches below the monument stone itself. We proceeded to tie out these stones and then remove them.

The first indicator that something was wrong was the concrete base that this original stone was sitting in. I know for sure they didn't haul redi-mix up to the south rim of the Grand Canyon in 1910 when these monuments were set! Now I knew that someone had been messing with these monuments at least to some extent.

The only place that the bearing trees fit the stone was at corner number 1. As you proceeded around the corners through the HES they continued to get further and further away from the trees as compared to the what the record called for. Under each stone was supposed to be a memorial stone. I could not locate the memorial stone under any of those stones except corner number 1.

The next step therefore was to use distance-distance intersections from the bearing trees and dig down there. At most of the other eleven locations I found the memorial stone at that distance-distance intersection. Now I had proof positive that someone had moved the original stones to a different location for some reason.

I sat on this project for three months because I was very concerned about the impact it would have in that small community. Virtually every piece of land sold or leased within that hundred and fifty eight acres was tied to one of these stones. Here I was about to tell them that only one of those twelve stones was in its original location and that someone had moved them. Further complicating the matter was the fact that many of my friends who had done surveys in the area over the previous years had used those very stones. I knew I was going to really upset the apple cart in this small community of Tusayan.

After some soul searching I realized the only ethical thing to do was to do the survey right. I completed the survey by setting my own monuments at the proper locations. I left the stones in the positions in which they had been discovered simply because I was concerned about the number of other surveys that had tied to those locations. I filed a multi-page plat which said specifically and in great detail what I had done, why I had done it and what I theorized had occurred. But now you need to know the rest of the story.

Some four years later I was at the New Mexico Surveyors and Mappers convention in Albuquerque. I was sitting at a table in the bar and over heard two surveyors discussing a survey that one of them had done back in the fifties at the Grand Canyon. He explained enough things to this third surveyor that I realized that the gentleman sitting at the table next to me had been at the Grand Canyon at the time of the survey of which I was suspicious.

When their conversation broke up I started a conversation with the first surveyor and asked a few detailed questions, And sure enough, I verified that he had been there with that crew. I asked him if they had moved those stones and he said, "Of course we did, they didn't match the record". I asked, "How do you know they didn't match the record? Did you check the bearing trees? He said, "We didn't know anything about bearing trees at that time". "So what record were you trying to match" I asked. His response was, "The bearing and distances around the HES".

Apparently what this crew had done was take a solar observation at corner number 1. They accepted corner number 1 in it's actual position. They proceeded to run perfect record bearing and distance around the HES and of course the further they went along the survey the farther it got away from the monumented stones. So they proceeded to pick up the stones, replant them in the same orientation but in a new location that fit the record boundary.

I can understand a person who has little or no experience in surveying doing these sort of things. With all due credit, the surveyor who openly told me what was going on was not licensed at the time and did not really know what was going on. I appreciated his honesty. But some registrant did sign a plat on that survey and that registrant to this day feels that he did nothing wrong.

I'll leave it to you, "Believe it or Not".

In the fourth story:

A registered surveyor had a contract with a Federal Agency. The surveyor was required to base all of his work on a solar observation. Although he had claimed in his proposal that he knew all about solars, in fact he had never made a solar observation in his career.

The government's representative for the contract decided to go out and show him how to take a solar so they proceeded to the field at a test site. They were going to take several solars and get this surveyor trained so that he could do solars all over his project.

The government surveyor asked the contractor to level his gun (instrument) while he got the calculator ready. The contract surveyor completed the leveling of the instrument in about five seconds. This concerned the government representative and so he went over to look at the gun. He noticed that the contractor had set the gun up, leveled it with the circle-bubble on the tribrach but did not turn the instrument in any way to check level in any other direction with the vial.

Here was a registered, professional surveyor who did not know how to level your standard surveying instrument! The government surveyor was very gracious by instructing the contractor in a more thorough and complete method of leveling the gun.

They got to taking solars a few minutes later.

The fifth story:

It begins with a friend of mine in the great state of Washington telling me about an event. He's a county surveyor and part of his tasks include the review of other peoples' surveys. In this particular instance he had opportunity to review the plat of another registered surveyor. The plat had several strange indicators on it that caused him to ask for a copy of the fellows field notes.

In reviewing the field notes of the traverse the county surveyor noticed that the registrant had not turned the final angle or measured the closing distance of his closed traverse. He called the surveyor to discuss why this had not been done.

The registered surveyor proceeded to explain that there is no need to turn the angle or measure the distance of the final leg of a traverse. All one has to do, he said, is set up coordinates on the traverse and inverse that last point. Then when he ran a closure on the survey it always closed perfectly.

The surveyor was dead serious! Through twelve years of registered experience he had never closed a traverse and honestly thought he had perfect closures all along!

The sixth and last story involves the New Mexico state land office:

In 1990 there were no surveyors employed. I had done considerable training in the area of legal descriptions and basic land law for their staff.

So when they came upon a lawsuit involving surveying they needed an expert witness. It was natural that they would call me.

The law suit involved the location of a house down near Las Cruces, NM. The question was whether the house was on State land or on private land. The state of New Mexico was claiming that it was on state land and was therefore trespassing. The owner of the structure of course felt that it was on private land and was wanting to fight this through the court system. Much of the preliminary proceedings had already taken place and it was decided that I would be deposed by the land owner's attorney. It was also decided that the state attorney generals office would depose the land owner's surveyor.

The land owner's surveyor had been in the surveying business for thirty years in southern New Mexico. His name is on many a plat, his number on many a monument throughout that part of the state.

The deposition took place via telephone. The land owner's attorney asked me three or four questions most of which were irrelevant to the real issues of the case. I could tell that he knew absolutely nothing of what we were talking about and further that he was getting very poor advice, if any, from their surveyor.

The deposition ended with very little information coming out from me. But then in the same telephone conversation the deposition was to be made of the other surveyor. I had spent two hours with the state attorney generals office giving them questions and lines of discussion that would net the information that we wanted. They proceeded to do an excellent job in deposing that surveyor. And on that day that surveyor disgraced our entire profession.

When asked how he knew if the house was on private land or state land his response was that a title company had told him. Several other answers that he gave indicated to me that perhaps he had never been on the site. I wrote a note to the state's attorney to ask that specific question and when she did his response was, no, I've never been on the site, I don't need to be on the site. The deposition went down hill from there. He was asked, "did you look for any monuments on the ground?" His answer was no! They asked why he didn't do that and he said, "because all of those corners are lost."

He was then asked how he know they were lost. Again his response was, "The title company told me. I rely strongly on what the title company says". He was asked if he thought it would have been valuable for him to go out and search for the corners himself. His reply was, "Well, I could go out there and look but I really wouldn't know what I'm looking for, you never know what they set." The states attorney immediately asked him, without any prompting from me, "do you have a copy of the GLO field notes?" The surveyor responded, "No, those are not available to the public and never have been".

Hard to believe, isn't it? The man was either a liar or totally incompetent. I believe he's both. I did not leave that session feeling proud of myself or of my profession. I was deeply ashamed of how my profession had represented itself to that land owner and his attorney.

Interestingly, about three hours later I received a call from the landowner's attorney. He asked me a couple of questions. I was very careful in my response since we were no longer in a deposition situation. But he finally came right down to his question. He asked, "what do think of our surveyor?" I couldn't hold back. I'd had a few hours to think and I told him this, "That surveyor is an absolute embarrassment to our entire profession. Based on what he said here he ought to have his license revoked and he ought to be punished severely".

I really hoped this case would have gone to court because I would have loved to have seen this ridiculous scenario played out to it's fullest. But it did not. The landowner's attorney wisely dropped the suit and backed off completely. I only wished that the events of that day could have been broadcast to all of the citizens of the state of New Mexico. That surveyor doesn't deserve to earn another dime in my profession!

Chapter 49

THE ETHICS QUESTIONNAIRE

Starting in 1989 this author (with the help of Roger Green of Oregon), conducted a number of ethics seminars for the state surveying associations. The questionnaire located in Appendix B was given to 2,398 persons attending those meetings. Mr. Green and I designed the questionnaire to try to determine what surveyors thought of ethical issues.

I will be the first to admit the questionnaire has a few shortcomings in it. Some of the questions may not have been as clear as they could have been. And some may have not been relevant to certain persons in different applications of the surveying profession. However, this "poll" is an excellent tool and represents a significant sampling of the profession.

Further, the manner in which the rating of ethical issues was presented caused some persons a little heartburn. I apologize for that. But I also know you can never write a perfect questionnaire for a group as diverse as the surveyors of the United States. In the end, the surveying profession was able to assess what is ethical behavior and what was not.

The questionnaires themselves were assessed by myself over a four year period to see where the profession stands. I think you will find the results interesting. I have not addressed all the questions from the questionnaire in this chapter, but have selected those I felt were most telling.

BACKGROUND INFORMATION

Of the 2,398 respondents, 77% were licensed as surveyors, with 8% being PE's and another 6% as LSIT's. The remaining were "other", presumably surveying technicians. A full 100% were high school graduates with 71% claiming some amount of college. 48% had received a BS or BA.

The average age of the respondents was 38 years, with an average of 14 years in the profession. Of those participating in this poll, only 7% were members of ACSM. State association membership was at 89%, but one should remember this poll was taken at state sponsored meetings. Obviously that statistic is severely warped! Only 3% had ever seen the NSPS *Surveyor's Creed and Canons.* (See Appendix A.)

ETHICS ISSUES

We set up a rating system for the first 25 questions (see Appendix. B). I want to highlight a few of the issues raised which seemed to be sending a message (either good or bad) to us about our view of ethical issues within our profession.

The following issues were rated as "very ethical" things to do:

> Attending continuing education courses
> Taking a loss on a job so you can do it right

A few issues rated as "OK" or "ethical" that were surprising were:

> Laying out property lines without regard to lines of occupation
> Turning in a surveyor to the Board without talking to him/her first
> Setting an additional monument 0.30 feet from another.
> Performing a survey without benefit of a title report.
> Rejecting/accepting monuments based on "who set it".
> Running out GLO record after adjusting the misclosure.
> Calling a monument out of position but not setting a new one.
> Side shots to points without using independent checks.
> Moonlighting without your employers knowledge
> Moonlighting with your employers knowledge
> Laying out property lines based on occupation only
> Performing a survey without benefit of researching the record
> Not looking for original evidence because all adjoiners accept the fence lines

The issues which generally were rated out as "unethical" were as follows:

> Destroying another surveyors monument
> Giving a good recommendation to the Board for an undesirable applicant
> Moving a monument 0.10 feet to agree with the record
> Failure to mark a monument with your registration number
> Working outside your area of expertise
> Surveys being "rubber stamped"
> Withholding survey information from other registrants
> Not informing a client of a possible boundary conflict

An in-depth analysis of each of these could prove interesting, or exhausting. The majority of these issues are discussed elsewhere in this book. It is interesting to notice the types of issues in each category.

The unethical list seems to be more "black and white" type issues. The "OK" issues tend to be more abstract issues involving the real nuts and bolts of professional surveying. Some of them are shocking given the right background of the question.

ADDITIONAL QUESTIONS

At the end of the questionnaire we had some other questions for respondents. Many of their thoughts were incorporated into this book. However, here are a few highlights worth mentioning.

A whopping 92% of those polled said they knew of a registrant who was operating in an illegal way! Their reactions to that registrant are as follows:

Talk to them	32%
Turn them in	3%
Spread word	8%

The number one reason cited for not turning in a bad practitioner was the "fear of reprisal". That is understandable, since most surveyors commented that they too "probably had some errors out there that someone would use against them in revenge". Some others cited the fact that they have to "work and live in the same community as these others, and could not afford to stir that pot".

Interestingly, 83% felt that doing a slipshod survey was a form of lying to their client, and 79% saw it as lying to the public. When asked how we should deal with those who regularly perform poor work, only 30% felt they should be turned in to the Board. These anomalies defy explanation.

Only 16% of those polled felt surveyors should be required to have their work reviewed by another professional before final approval. Some commented that they already have this through county surveyors, but most felt the review by those officials was bureaucratic, irrelevant, or self-justifying.

About 28% of those polled felt the wages they pay their employees was an ethical issue. Most felt it was purely a matter of economics, supply and demand, or level of expertise.

The most interesting question was the one asking if the respondent had ever done an "illegal" survey. The question leaves one wondering if it was intentional or not, and this doubt may have skewed responses. However, 37% felt they had performed an illegal survey, however they interpreted the question.

Finally, a question was asked on whether or not you would perform a survey differently based on three different givens. The responses were as follows:

Who the client is	11%
The fee you charge	12%
Who the adjoiner is	27%

The vast majority of us appear to feel the changing of a survey procedure/result would be rare based on who or how much (although 11% and 12% are surprisingly high to this author). The interesting issue is the one about who the adjoiner would be. Is it possible that more than one fourth of our profession would change their survey based on who owned land on the other side of the line? That is worth thinking about.

CONCLUSION

Most polls are somewhat inconclusive. They provide us with a snapshot idea of where people are. Things change, so do ethics. It is important to remember that the respondents to this questionnaire are the "cream of the crop". They are people who are involved in the profession to some extent, given their attendance at a state association sponsored event. The real unknown is that large group out there who does not attend any activities and belongs to nothing except the human race.

The information presented in this chapter is for your review and benefit. There is no claim that it is an absolute about our profession. But it does give us an idea of where some of us might be. If you were surprised about some of the responses, you would probably be shocked at responses from those who are "independent surveyors", who never participate in the profession.

I want to thank all those who participated in the sessions that helped us gather this data. It helped me immensely in the content of several of the chapters.

Chapter 50

RESEARCHING THE RECORD

> "A resurvey, properly considered, is but a retracing, with a view to determine and establish lines and boundaries of an original survey,...but the principle of retracing has been frequently departed from, where a resurvey (so called) has been made and new lines and boundaries have often been introduced, mischievously conflciting with the old..."
>
> *Cragin v. Powell (128 U.S. 691 , 1888)   U.S. Supreme Court decision*

In several locations in this book, you will find my reference to the failure of some surveyors to research the record. What is meant by the record is referring to the federal, state and local records that would affect the outcome of a survey. These records would include the GLO/BLM survey notes (if applicable) as well as the records of other surveyors who have worked in private practice or for government entities.

My experiences in surveying have included working for several organizations which never research the record. It seems their chief corner search tool was the USGS quad sheet map. As most surveyors should know, the USGS map is a useful tool, but it is extremely limited. The quad sheet cannot tell you what the corner monument is or what accessories set. It also cannot tell you if any remonumentations have taken place by the public or private sector. The USGS cannot provide data on who has used that monument before nor in what process it was used.

I'll be the first to admit that records search can be complicated, time consuming and expensive. In reality, it may be the most expensive part of a project. As time goes on, that would certainly be the case; records research would be more expensive. Researching the record (the public record) is far more difficult in some states than in others. I applaud states which have had mandatory plat recording laws for long periods of time. They have forced the private sector to place information crucial to public welfare into a public repository. But the majority of states are lacking in this regard. In fact, some states have absolutely no public record to speak of.

I am reminded of the state of New Mexico, where I worked for nine years. That state does allow the surveyor to file a plat. But many counties have never had a plat filed. In Bernalillo County (which includes the Albuquerque area) there are several hundred survey plats recorded with the county. However, the indexing system requires you to know the name of the land owner or the name of the surveyor. It is not indexed by any geographic system. So researching the record in that county was almost a joke. It's no wonder that some surveyors are very frustrated with records research. When you have to deal with these sort of problems, it can be expensive and perhaps even fruitless.

In spite of the difficulties that one can encounter in records research, there is nothing more fundamental to the purpose of licensing surveyors. We are a regulated profession because the society wants some guarantees from the state that we are going to protect their welfare and safety. How can you do a survey without researching the record? How can a surveyor walk in the footsteps of the surveyor before him, unless he knows what those footprints look like?

There are tremendous advantages to having a vast record to research.

- Your liability will greatly decrease. As the surveyor has access to a more complete history of every corner point and area, he can make far more professional decisions about the results of his survey.

- The surveyor who is constantly being placed in the situation of having to provide estimates for a survey, can gather tremendous data from the record.

I remember a time in Arizona, when I had to bid on a job with the National Park Service. It was to perform a survey of the Tuzigoot National Monument in central Arizona. The Park Service had contacted several local surveying firms; mine included. We were given a week to provide a solid price estimate (or bid) on doing a boundary survey of the project. Arizona has a slightly better than average records system. So I went to the county courthouse to look and see if anyone had recorded any surveys in the area. I was pleased to find that all three sections involved in this Park Service project, had recently been surveyed by another Federal agency. I reviewed their plats and found that everything they had done was excellent quality work. Based on that information, I put in a price to do the boundary survey for about $10,000.

Three weeks later, I was awarded the project. As word spread through the local area that I had got the job, my competitors started to call and complain about how I had "low balled" the price. They were quite open with what they bid and most of them were in the $20,000 range. I always hate to leave too much money on the table, but I felt comfortable with my price. Not one of my competitors was aware that this other survey had been done. They had all assumed that they would have to start form scratch in the project area.

I completed the project in five working days. I made a profit and I felt very comfortable with the price and the product that I provided. They say that "Knowledge is Power". That is certainly true. Some day the surveying profession will have to wake up to that fact, and admit that it would work in surveying, too. The more information you have, the more valuable you will be; the wiser you will be in your price proposals and other project negotiations. Sharing information has worked for many other professions. Consider the realtors with their multiple listing service. How about the real estate appraiser who in most states has access to a vast data base that allows him to search for comparable sales quickly and efficiently. Consider the benefits of having a governmental entity that records deeds and other related documents. What chaos would we face if there were no official public records on land ownership? Sounds almost ridiculous. Survey records are exactly the same.

If you reside in a state with little or no survey recording requirements, start a campaign with your association to make the change. Such legislation needs to have a geographic indexing component to it. And it needs to be mandatory; all surveys must be recorded. Let's bring our profession into the real world, the "information superhighway". The only ethical thing we can do is provide the public service we are licensed to provide.

Chapter 51

CORPORATE RESPONSIBILITY

> "People are not born to inaction; they learn it."
> *Charles Garfield*

America is the land of the corporation. We have literally formed millions of legal "entities" which serve our needs in liability, taxation, and organizational structure. The climate for business in the United States has usually been favorable. It is part of what has made us a great nation. I do not wish to argue the pros and cons of business climates, nor the best type of structure to form. The subject for this chapter is not about S Corporations vs. C Corps. It does not matter whether you have incorporated or not. This subject is about your responsibility as a business in this great land.

Even the poorest persons in the United States live like kings when compared to the average citizen of this planet. We are truly a blessed nation. Most of the other western nations have similar standards of living. While we often complain about hard times, we have lost sight of what the rest of the Earth must deal with on a daily basis. I am reminded of the words of a country song from a few years back: "We'll bitch about a dollar, when there's those without a dime." We need to keep an accurate perspective of where we really are.

We in the wealthy western nations have done well because of the benefits of our legal and taxation systems. We've also done well because of our "work ethic". This ethic can make one wealthy, but it can also take one down a path that destroys very important elements of life, like relationships and true purpose in life. But whatever the reasons, we are generally doing very well compared to others.

Business has an ethical responsibility to return some of it's "blessing" back to the society that helped to create it. This can be done in both a micro and macro environment. But there is something to the concept of "it is more blessed to give than to receive". Corporate America has sometimes lost sight of this principle. And just like the unwritten law of gravity, you cannot break this principle of giving back. It will always come get you when you violate this concept.

I have watched it for years. I have seen many a business (not just a surveying business) which has reaped huge profits from society. I do not say they did not deserve their profits. Obviously their products or services were of consistent need and use of the public. But after a few years, these organizations which earned their profits from the people will forget how and why they are so well off. They will get greedy. They will become arrogant. And they often will trample on the very people who helped them attain their initial status. It is short-sighted and just plain foolish. Yes, there is a Corporate Responsibility to give something back.

There are a number of ways a business could "return the favor". A very basic method is to simply donate cash or services to organizations in need of help. While some "charities" are just concealed cash-producing machines, most are legitimate agencies helping in an incredible array of services and aids. We cannot let the presence of a few charlatans get in the way. Do a little research on potential recipients of your generosity if this is a concern. From organizations that exist to serve the homeless to those assisting aspiring students, there are many good opportunities. What about donations for medical research, especially in an area in which one of your employee's mates is suffering? Or perhaps a few dollars to a young aspiring child of an employee?

Cash is the simplest way to give. Often it is the most needed. These organizations can often stretch a few dollars farther than you and I can imagine. They are good at it, because they care. And you should care too.

Services are often ways to provide a donation, especially if you do not have cash to give. I once did a full survey of a parcel of land for our volunteer fire department. They were planning a new fire station. As usual, the department was cash poor. The survey would have cost them about three thousand dollars. I charged them $250 to cover the wages of my helper; the rest I donated. When you think about it, there are many such opportunities out there. I've been proud to help raise over $50,000 for surveying scholarships over the years. I just donated my "services" of giving a seminar to do it.

What about donating your employee's time (you cover wages) for a few hours to work at some community project. There's "Habitat for Humanity" that builds homes for low-income persons. Your employees could work at the simplest of tasks for a soup-kitchen at Thanksgiving, or perhaps help plant trees and shrubs for an improvement at some park. I've seen a few places where surveying and engineering firms have participated in an "adopt a highway" clean-up project. I know one firm which purchases toys for disadvantaged kids every year. And I know another which donates their employees for a day to deliver coats before winter arrives. Why not sponsor that little league team? There are an unlimited number of ways to give like this!

Another way to give back to your community is to take good care of your employees who have helped you make that money. This should be more than a Christmas turkey. You can give to them in many ways which do not cost you much money. Allowing them a flexible schedule to participate in their own charitable efforts is one major way. You could also provide a free lunch to them when they are cleaning up that highway under your corporate name. Or perhaps you could provide better facilities at your office for breaks, entertainment, or other beneficial activities. Many of these things "buy" tremendous goodwill toward the business. Ask yourself these questions:

Is my business an active part of my community?

Does the community know me as a professional, or as a part of the community? Or both?

Have you really given back to those who helped you earn your profit?

Do you feel a sense of responsibility to society as a whole?

That last question is where this discussion really comes down to the "brass tacks". The ethical issue here is how you feel about your profit. Is it "yours"? Or is it a reflection of many things, not the least of which is the providence shown you in this life? Oh yes, it was your hard work, effort, determination, and ingenuity. But let's not be so arrogant to think that our successes are based solely on these self-appointed attributes. You have an ethical responsibility to give back. We must be aware of our role as business leaders. Business leaders don't just make money — they lead their communities to bigger and better successes. Those successes are not always measured by profits.

So what kind of a person are you? Not what kind of business wizard are you....but what kind of human being are you? A person's true wealth is measured not by his income, but rather by his generosity. Don't forget the who, what, and why of your success. When you do forget, that gravity-like principle will cause you to hit the ground hard. Yes Corporate America, you have an ethical responsibility to give back. The surveying profession should become the leaders in this arena! It requires action!. We have learned to sit back and "let others take care of it." This is not acceptable in a land as wealthy as this. What kind of person are you?

Please find ways to be generous. There are many deserving students, people, and organizations. Go find them. Give freely and without complaining. And then sit back and know that you did not just "take" from this life. Rather, you were a part of great principle at work in a mighty nation. You were a *giver*. There can be no greater title for a truly ethical professional: Giver.

And on behalf of the rest of society, I thank you for your generosity.

Chapter 52

SERVING ON THE BOARD

"Responsibilities to the public demand that the surveyor place service to mankind above personal gain."

Andrew L. Harbin

In every state there exists the equivalent to a "Board of Registration" for Land Surveyors. Often these Boards also register other disciplines, especially Engineers or Architects. The Boards generally are authorized to screen, test, and regulate the professions under their charge. The extent of these actions, especially the regulation part, vary considerably from state to state. Overall, the trend seems to be more regulation and monitoring in most states.

An honor I have never had is that of serving the profession by appointment to a Board of Registration. To serve our profession in any capacity should be an honor, and this is perhaps one of the more meaningful ways one could serve. In most jurisdictions, this is an appointment by the Office of the Governor. However the appointments are made, it causes one to suddenly become involved in the very heart of the Land Surveying profession.

Today, Boards spend more time on disciplinary actions than the screening and testing process. I suppose that is a sign of the times. Frankly, it's about time we started getting a lot tougher on those who have violated our public trust. Is this not what the public has granted to you and I as registrants...a trust that we know what we are doing and we do it right? Serving on a Board cannot be a completely enjoyable effort. But who said service would be a "walk in the park"?

I would like to offer a thanks to all those who have served on the Boards of Registration over the years. Your service has usually gone unnoticed. And it is easy to take "potshots" at the Board. Those of us on the outside rarely know all the issues, facts, or complexities in any situation. We need to give the Boards our support and encouragement.

So, thanks for your service folks!

It is important that those who serve, or who will someday serve on a Board consider the extra ethical issues they must consider. Being on a Board can be a very powerful and influential position. And it can be abused. Board members are not just concerned with the ethics of the applicants and registrants. The ethical implications of serving on the Board are very significant. It would be good for us to briefly review some of the potential ethical issues for those who sit on Boards of Registration.

The most common and obvious source of ethical problems for the Board member is that of the "conflict of interest". We should also expand this concern to the appearance of a conflict of interest, which usually goes farther into the gray area. Conflicts of interest can arise whenever a Board member has had some relationship or encounter with an applicant or registrant outside of the Board service. This includes the appearance of helping your own employees, friends, or co-workers get through the "process". The wise member will consider abdicating their office for these type considerations. No matter how impartial you really are, appearances of conflicts are not worth trying to conquer.

An especially difficult situation is when a registrant is brought before the Board over some alleged improper action in a survey. If a Board member has had some previous dealings with the registrant (especially in a negative way), there can be a real cloud hang over any judgment. Perhaps the "accused" registrant once disagreed with the Board member on some other work five years ago. Or was there some "bad blood" between the two individuals over some other issue in the past, perhaps over one "stealing" a client from the other? Impartiality often disappears in the eyes of others when we get ourselves into these situations. The best thing to do is announce immediately that there may be some conflict and you wish to step aside for this particular hearing. This should be done in a way which does not further prejudice the remaining members of the Board!

I have witnessed Board members who abused their power. It has happened where they have overly punished someone who is their competitor. I've seen service on the Board used in advertising and marketing to take an advantage over the competition. This is a blatant misuse of the opportunity to serve the profession. I've also seen Board members whose appointments were solely for political reasons; there was no consideration of their real knowledge level or ability to comprehend complex legal issues. They were quite ignorant of some basic surveying principles, but because they had the "mantle of office" they lorded over us in spite of their inexperience and ignorance. Like an arrogant little Napoleon, they wanted to be sure we knew they had the power.

I once found a major error in a Board members survey. It was an obvious lie on a plat and to his client. When I brought this to him in private, he flatly refused to fix it. He admitted it was wrong, but would not change it as it would "cost him too much". When I suggested this was a violation of the regulations in that state, he smiled and asked me,

"Who do you think would rule on your complaint to the Board?" I think there was something wrong with that picture.

Being in a public office such as serving on a Board usually brings one under closer scrutiny than normal. It behooves the Board member to mind his or her "p's and q's" in many settings. The Board member who innocently or purposely leaks sensitive information to outsiders has violated the office. I've heard things I should never have been privy to simply by sitting at the same table as a Board member for lunch. Names, issues, and accusations have been exposed by loose lips of Board members at survey conventions and other social settings. Please be careful of this trust we have given to you.

This issue of "closer scrutiny is similar to that of political offices. While not as intense, the similarities are worth considering. A would-be presidential candidate should consider what their background, and current situations can do to their chances. The media will find it out and eat them alive. They are foolish to have thought they could get by with these issues not coming to light. Similarly, the Board member must realize that their personal habits and attitudes can cast a shadow over their ability to serve properly on the Board. For instance, the Board member who regularly gets drunk at the convention is an embarrassment to the profession and to the credibility of the entire Board. As mentioned in the first chapter of this book, our personal ethics are difficult to separate from our professional ethics. This is especially true and important when serving in a public office of trust.

You will be held to a higher standard....and you should be! That goes with the office. Board members are literally the most prominent members of our profession. They need to set the right example. They need to be leaders, in every sense of the word. Certainly, ethical issues should be of great interest to the Board member in a very personal way.

One last issue to mention is that of the role of "judge" over the ethics of another professional. Frequently, the modern-day Board member finds him or herself in this situation. Registrants are being called in to explain their actions to Boards, both technically and professionally. While I am pleased to see the Boards acting in this role, I offer a couple thoughts of caution. The truly wise and circumspect Board member will have two very important tools at their disposal. These are:

- Be sure you really know the rules, laws, and principles you are judging. This will require you to attend conferences, seminars, and other opportunities to be constantly "sharpened" in these areas.

- Be sure you recognize a gray area and take that into consideration when you deal with the problem. The most dangerous people in

Land Surveying are those who think everything is black and white. The Board member should be the expert on what areas are gray, and which are not. Know when to get outside help....it is the sign of a true professional.

I've never been called before the Board for a mistake.....at least not yet. I hope I never do, but I'm just as human as the next guy. I have testified before Boards, served on committees, and advised them on technical issues. I have witnessed Boards go through some very difficult situations. And I have been in awe of the lengths to which most Boards go to make things right. I have the utmost respect for the Boards and the majority of people I've known who served on them. Thanks again for all your effort. If you keep a proper and cautious attitude about ethical issues, you will continue to serve us well.

Chapter 53

POTPOURRI

> "Who's on first?"
> *Abbot and Costello comedy routine*

In this book you will find a chapter entitled "Believe it or not". It is a collection of stories that I have heard or experienced which I found hard to believe, but were true.

This short chapter is a potpourri collection of one line statements I've heard from registered professionals. Some settings were in meetings, some in court, and some in general conversation. I could probably write a chapter on each of them. But I wanted to share a few of these with you to simply get you to thinking about where our profession may be or where it is headed.

So fasten your seat belt, take some anti-nausea medication, and read on. Although severely tempted, I will not comment on these. You are left to your own conclusions.

1. Surveyors should never concern themselves with anything other than what people are occupying.

2. Meander lines are always fixed boundaries.

3. The GLO record is not available to surveyors.

4. Original evidence is too expensive to look for, and a waste of the clients money.

5. Now that we have EDM, we should rely on distances more than evidence.

6. After 30 years in the profession, I've found I really do know everything I need to know.

7. If we just shot all the lawyers we wouldn't have any boundary disputes.

8. Intent of a deed should not be limited to what the deed says.

9. Bearing trees are always marked like "this".

10. Only the government can afford to look for evidence. The rest of us just do what we can.

11. The USGS quadrangle is a very accurate source of knowing whether a corner is in or not.

12. Section lines are always straight between a section corner and a quarter corner.

13. Local custom should outweigh the BLM Manual as to how to reset a lost corner.

14. In a metes and bounds description, the metes are far more important than the bounds.

15. The less information I put on a plat the better. That protects my client and myself.

16. You only have to turn the angle once, as this gun reads to the second.

17. Strength of figure is no longer an issue with the very accurate instruments we now use.

18. I don't care if the existing monument is only .01 foot out of position, it cannot be accepted unless it is perfect.

19. Surveyors have no interest in adjoining deeds; my client gets exactly what his deed says because it is a warranty deed.

20. If the client is not willing to pay for a full legal survey, I will provide whatever service he can afford.

The truly ethical and professional surveyor will be well informed as to the realities of working in this complex profession. The above statements are all real comments I've heard over the years. They indicate that some do not know some basics of our profession. The problem is spread from the instrumentation to the legal principles. We have a problem! Is it possible that licensed individuals do not really grasp Surveying 101? Yes, and you find proof of that fact every time you go to the field.

We as a profession cannot allow this to continue. I do not accuse these people of intentionally doing things improperly, although I'm sure some have. I am more concerned that we as a profession deal with this rather than have others deal with it for us. Or worse yet, our profession could actually disappear if we do not grab hold of these misperceptions and turn them around.

The only solutions I know of are education, disciplinary actions, and revocation of licenses. Let's get this together, shall we?

One more comment; this from a registered surveyor who is also a GIS "guru" in a county GIS office in the southwestern U.S:

> "Once we get a position on every property corner, we really will no longer need any land surveyors."

My fellow Professional Surveyors, do we really want others to determine our destiny as a profession? If not, then we better be about that business ourselves. And let's not delay—that GIS "guru" may be your neighbor or boss someday. He may have the ear of a State Senator or Council Member. We must take control of our profession by eliminating those in our own profession who cannot fathom why they are surveyors. And we must educate those who "think" they are surveyors, but who only know a vague mathematical shadow of what real land surveying is all about.

The public's welfare and trust are built into your surveyor license. The only ethical thing we can do is live up to that trust. We cannot afford a potpourri of measurers, computer jocks, amateurs, weekend warriors, and wanna-be's to operate under the term Professional Surveyor. That term should be reserved for the real thing. Please think on these things; our future depends on it.

Chapter 54

CONCLUSION

"There is no real excellence in all this world which can be seperated from right living."

David Starr Jordan

Congratulations on making it through this entire book. If you've made it this far I probably owe you a beer! The preceding chapters may seem to be random thoughts thrown at you. But I assure you, they were thought out carefully for many months. I have not given you too much of a list of do's and don't's. Rather, I hope I've planted some seeds for future events in your professional and personal life.

If you want to find a common thread in these issues, I believe it would be that of honesty. Almost every ethical issue somehow revolves around a sense of honesty. Whether it be your financial dealings, your measurements, or your words, honesty and ethics are regular bedfellows. If you are basically a dishonest person, you will struggle with ethical problems your entire life. Honesty requires a lifetime commitment. And it must be in all matters. You cannot be selective in how you dispense honesty. In fact, to attempt to be selective is in and of itself dishonest.

In the rear of this book you will find a reprint of the current "ethics" for our profession as outlined by ACSM/NSPS (See appendix "A"). I hope you will review them and meditate on their deep meanings. That list may not be complete. I know of some who disagree with the ACSM ethical canons, and I understand. But I truly believe they are a great place to start.

While we can complain about our profession as a whole over ethical issues, we cannot really deal with them at that level. It is the individual surveyor who must answer to him or herself. Where are you when it comes to ethics?

I love my profession. I really am glad to be a surveyor, and I'm grateful for all the opportunities I have had over the years to work and serve in this profession. This book certainly is not the epitome of this subject, nor of my career. But I deeply hope that the thoughts and ideas conveyed herein are of great value to you. I love this profession. Let's make it even better. I wish you all the best.

Appendix A

SURVEYOR'S CREED AND CANONS

As a Professional Surveyor, I dedicate my professional knowledge and skills to the advancement and betterment of human welfare.

I pledge:

To give the utmost of performance;

To participate in none but honest enterprise;

To live and work according to the laws of humankind and the highest standards of professional conduct;

To place service before profit, honor and standing of the profession before personal advantage, and the public welfare above all other considerations; In humility and with need for Divine Guidance, I make this pledge,

Canon 1:
A Professional Surveyor should refrain from conduct that is detrimental to the public.

Canon 2:
A Professional Surveyor should abide by the rules and regulations pertaining to the practice of surveying within the licensing jurisdiction.

Canon 3:
A Professional Surveyor should accept assignments only in one's area of professional competence and expertise.

Canon 4:
A Professional Surveyor should develop and communicate a professional analysis and opinion without bias or personal interest.

Canon 5:

A Professional Surveyor should maintain the confidential nature of the surveyor-client relationship

Canon 6:

A Professional Surveyor should use care to avoid advertising or solicitation that is misleading or otherwise contrary to the public interest.

Canon 7:

A Professional Surveyor should maintain professional integrity when dealing with members of other professions.

Appendix B

ETHICS QUESTIONNAIRE

This questionnaire is designed to assess your feelings and opinions on the subject of professional ethics. Obviously there are no right or wrong answers. Feel free to speak your mind and add comments where you want. Some of these topics will be covered in more detail in today's session.

BACKGROUND INFORMATION

1. What is your current job title?

2. Are you a registered LS?_____ PE_____ LSIT_____ Other_____

3. What level of schooling have you completed?

      High School_____   Misc. College_____   AA _____

      BA/BS_____   Masters_____   PHD _____

4. How old are you?_____

5. How long have you been in the surveying profession?_____

6. Are you a member of ACSM? Yes   No

7. Are you a member of your state surveying association? Yes   No

8. Have you ever seen the ACSM/NSPS code of ethics? Yes   No

ETHICS ISSUES

Please rate the following questions in this manner:

      0 = not an ethical issue in my mind   3 = OK
      1 = very unethical   4 = ethical
      2 = somewhat unethical   5 = very ethical

1. Performing a survey without benefit of researching the record._____

2. Laying out property lines without regard to lines of occupation._____

3. Turning in another surveyor to the board without talking to him/her first._____

4. Setting an additional monument within .3' of an existing one._____

5. Performing a survey without benefit of a title report.____

6. Rejecting/accepting monuments based on "who set it".____

7. Destroying another surveyor's monument.____

8. Not looking for original evidence because all adjoiners accept the fence lines.____

9. Giving a good recommendation to the board for an employee who was undesirable.___

10. Running out GLO record after adjusting the misclosure.____

11. Calling a monument "out of position", but not setting a new one.____

12. Failure to mark a monument with your LS number.____

13. Moving an existing monument a tenth to agree with the record.____

14. Attending continuing education courses.____

15. Taking a loss on a job so you can do it right.____

16. Side shots to points without using independent checks.____

17. Regular usage of an NGS baseline.____

18. Not informing a client of a possible boundary conflict.____

19. Working outside your area of expertise.____

20. Training your employees beyond their current needs.____

21. Moonlighting with your employers knowledge.____

22. Moonlighting without your employers knowledge.____

23. Surveys being "rubber stamped" by a registrant for a non-registrants use.____

24. Laying out property lines based on occupation only.____

25. Withholding survey information from other registrants.____

ADDITIONAL QUESTIONS

1. Have you ever known a registrant who was operating in an illegal way   Yes   No

1a. If so, did you talk to them____, turn them in____, spread the word about them____.

2. What are the reasons a surveyor will not turn in a bad practitioner?

.3. Do you think surveyors should be required to have their work reviewed by another registrant before it becomes final?

4. Do you see doing a slipshod survey as a form of lying to the client? Yes   No
   To the public? Yes   No

5. How should the profession deal with those who regularly perform poor work?

6. Do the wages you pay your people have any ethical implications?

7. Have you ever done an illegal survey?   Yes   No

8. Do you perform surveys differently based on who the client is? Yes   No
   Based on the fee you will charge?   Yes   No
   Based on who the adjoiner(s) might be?   Yes   No

9. What is your personal definition of ethics?

10. Additional comments (Use reverse as needed):

Thank you for your help in this important subject!